Flexible Bioelectronics with Power Autonomous Sensing and Data Analytics

Sameer Sonkusale • Maryam Shojaei Baghini •
Shuchin Aeron

Flexible Bioelectronics with Power Autonomous Sensing and Data Analytics

 Springer

Sameer Sonkusale
Tufts University
Medford, MA, USA

Shuchin Aeron
Tufts University
Medford, MA, USA

Maryam Shojaei Baghini
Indian Institute of Technology Bombay
Mumbai, India

ISBN 978-3-030-98537-0   ISBN 978-3-030-98538-7   (eBook)
https://doi.org/10.1007/978-3-030-98538-7

© The Editor(s) (if applicable) and The Author(s), under exclusive license to Springer Nature Switzerland AG 2022

This work is subject to copyright. All rights are solely and exclusively licensed by the Publisher, whether the whole or part of the material is concerned, specifically the rights of translation, reprinting, reuse of illustrations, recitation, broadcasting, reproduction on microfilms or in any other physical way, and transmission or information storage and retrieval, electronic adaptation, computer software, or by similar or dissimilar methodology now known or hereafter developed.

The use of general descriptive names, registered names, trademarks, service marks, etc. in this publication does not imply, even in the absence of a specific statement, that such names are exempt from the relevant protective laws and regulations and therefore free for general use.

The publisher, the authors and the editors are safe to assume that the advice and information in this book are believed to be true and accurate at the date of publication. Neither the publisher nor the authors or the editors give a warranty, expressed or implied, with respect to the material contained herein or for any errors or omissions that may have been made. The publisher remains neutral with regard to jurisdictional claims in published maps and institutional affiliations.

This Springer imprint is published by the registered company Springer Nature Switzerland AG
The registered company address is: Gewerbestrasse 11, 6330 Cham, Switzerland

Preface: Introduction and Overview of Flexible Bioelectronics

Motivation for This Book

Advances in materials, processing, and microfabrication have spurred the digital revolution of today. These advances have revolutionized the way we process and communicate information using computers and smart phones. They have changed the way we entertain ourselves using television and video games. And importantly, they have greatly improved the quality of our lives through advances in healthcare and assistive technologies. However, majority of these platforms still rely on designs and materials that are rigid and bulky. They may be well suited for the environment around us; however, they are ill equipped to adorn our body as wearables or be utilized as implantable, which requires conformability and flexibility. The need for flexibility brings new challenges in their fabrication that conventional microfabrication approaches that rely on top down lithography or high temperature processing using hard substrates may not be able to provide without further advancements. This requires new thinking and approaches to process soft materials into flexible devices that can then be integrated into systems for wearable, implantable, or other applications. Over the last decade, there has been tremendous research progress in this emerging area of flexible electronics. The research progress has spanned areas of innovative materials development such as soft polymers, including biomaterials as substrates, to the use of exciting two-dimensional ultra-thin nanomaterials such as graphene as transistors and sensors. Research has also focused on converting existing bulky and rigid materials like silicon into thin micro- and nano- membranes to make them flexible and conformal for integration onto flexible substrates. Age-old methods of roll-to-roll printing and inkjet printing have been transformed to allow fabrication of micron-scale features for flexible electronics. New device developments for transistors and sensors that are inherently more suitable for flexible devices such as electrolytic gating of transistors have also evolved. Finally, there have been a lot of peripheral developments in supporting technologies to interface and communicate with these new flexible devices. These developments may not be flexible themselves but are extremely important for practical realization of integrated flexible systems. For example, packaging and integration of flexible sensors with conventional Complementary Metal Oxide Semiconductor (CMOS) integrated circuits is key. Another example is that of wireless energy and power harvesting for flexible platforms, which may not be on flexible substrates. Yet another effort is on algorithms and circuits that are more amenable for processing data from flexible devices. These topics while not directly related to fabrication and realization are also quite relevant when it comes to the practical realization of flexible electronic platforms.

Flexible Bioelectronics

Flexible electronics in general has been touted as the next frontiers for consumer electronic devices like flexible displays, curvilinear televisions, and folding phones. However, the biggest driving force for advances in flexible electronics has been the need for devices that can provide intimate interface with the human body for monitoring our health and wellness. For example, there is a need to make next generation of prosthetics that can conform and interface well to the contours of the human body. Flexible bioelectronics is a niche area where flexible electronics overlaps with biomedical applications and where the key goals are to develop materials, devices, and systems that can reside safely on the human body or within it for the purpose of monitoring and augmenting human health and performance. Smart watches and smart rings are commercial products today that fit this space; however, these devices are still rigid and bulky and cannot be adapted for other biomedical needs. More promising examples of flexible bioelectronics take us beyond the smart watch to smart skin, where flexible devices conform directly on the human skin for health monitoring and provide intimate human–machine interfaces. Yet another example of flexible bio-electronics is that of smart bandages for the treatment of chronic wounds, where a biocompatible interface containing flexible sensors and drug delivery is needed to monitor and repair deep wounds. An example in the realm of implantables is that of a flexible neural electrode array for stimulation of and recording from neurons in the brain and peripheral nervous system. The next generation of neural interfaces also incorporates drug delivery and/or chemical stimulation for truly remarkable multi-modal neural implants. In such applications, there is an even more stringent requirement on the biocompatibility to ensure long-term chronic implantation. Every material, structure, or design used in the development of such flexible bioelectronic implants needs to be scrutinized for long-term toxicity and inflammation all the while ensuring continued functionality and performance over months and years of implantation. Different biomedical applications elicit different requirements for materials, design, and structure with unique processing and fabrication techniques more suitable for their form and structure. For example, neural implants are expected to have softness and mechanical properties of brain, which are more adequately achieved using hydrogels or soft biopolymers, which also happen to be more biocompatible. And these materials can only be patterned or shaped into devices using low temperature solution-based processing rather than conventional cleanroom techniques. These requirements are unique to flexible bioelectronics and form a subset of all approaches and methods available for flexible electronics. This emphasis will be the focus of the discussion in this book. It is also well understood that the progress in flexible bioelectronics will have to go through multiple phases of development, where the earliest and nearest developments will focus also on the practicality of merging simple concepts/devices for flexible sensors and electronics with conventional electronic devices that may be rigid but small (e.g., CMOS IC). These hybrid platforms combining small integrated circuit chips on flexible substrates through flexible interconnects can provide amazing flexibility and conformability adequate for many applications such as wearable electronic skins. The book will attempt to focus on such hybrid integrations, and also on conventional and novel circuits and algorithms that are suitable for such hybrid platforms.

Contents of This Book

Flexible electronics and bioelectronics has been a very active field of research and development in recent years. There have been many review papers written on this topic with select books. The focus of those papers and books has been solely from the point of view of material scientists who extol the virtues of polymers and flexible substrates, or those who focus on engineering hard materials through thinning and patterning to achieve flexibility. There are yet others that focus on specific applications. However, there is no discussion on flexible bioelectronics from the view of practicing engineers on how to apply their knowledge and expertise in devices, circuits, and algorithms to this emerging field. This book tries to address such a gap in understanding by first providing a gentle introduction to the materials and processing (Chap. 1) followed by basics in sensing (Chap. 2). Two detailed case studies of flexible bioelectronic platforms, namely smart flexible wound dressing and textile-based wearable diagnostics (Chap. 2), will provide a comprehensive review and deeper understanding of the challenges and promises of this field. Chapter 3 will finally delve into CMOS-based integrated circuits for readout from flexible sensor platforms covering both resistive and capacitive sensors. This discussion assumes a hybrid flexible bioelectronics approach where flexible sensors and flexible substrates interact with small CMOS ICs for readout and signal conditioning. Chapter 4 will continue the discussion of further digitizing signals using data converters amenable to sensing applications. Beyond circuits, efforts on energy harvesting and power management are critical for flexible bioelectronics platform and will be discussed in Chap. 5. Chapter 6 will finally discuss algorithms for processing and data analytics pertaining to flexible bioelectronic applications. Majority of the discussion will be on intelligent sampling and data acquisition to reduce power dissipation without sacrificing accuracy through compressed sensing approaches, and also data recovery and remediation in the presence of noise and motion artifacts.

Medford, MA, USA   Sameer Sonkusale
Mumbai, India   Maryam Shojaei Baghini
Medford, MA, USA   Shuchin Aeron

Acknowledgments

This book was a result of collaborative effort from DST (Department of Science and Technology) and MHRD (Ministry of Human Resource Development), Government of India, under the SPARC (Scheme for the Promotion of Academic and Research Collaboration). S.S also acknowledges the partial support of the Office of Naval Research (ONR) grant N0014-16-1-2550, ONR DURIP award N00014-20-1-2188, Department of Defense Congressionally Directed Medical Research Program (CDMRP) grant W81XWH-20-1-0589, Uniformed Services University/Henry Jackson Foundation (HJF) award HU0001-20-2-0014, and National Science Foundation (NSF) award 1931978, 1935555, and 1951104.

This book is a collaborative effort of multiple authors who contributed to the writing of this book. Below we acknowledge their contributions:

Preface
Sameer Sonkusale (Tufts University)
Chapter 1: Materials and Processing of Flexible Bioelectronics
Narendra Kumar (Boston College) and Sameer Sonkusale (Tufts University)
Chapter 2: Sensors and Platforms for Flexible Bioelectronics
Junfei Xia (Tufts University) and Sameer Sonkusale (Tufts University)
Chapter 3: Low-noise CMOS Signal Conditioning Circuits
Dr. Maryam Shojaei Baghini (IIT Bombay), Dr. Meraj Ahmad (IIT Bombay) and
    Dr. Shahid Mailk (IIT Delhi)
Chapter 4: Data Converters for Wearable Sensor Applications
Dr. Maryam Shojaei Baghini (IIT Bombay), Dr. Laxmeesha Somappa (IIT Bombay)
    and Dr. Shahid Malik (IIT Delhi)
Chapter 5: Power Management Circuits for Energy Harvesting
Dr. Maryam Shojaei Baghini (IIT Bombay), Dr. Meraj Ahmad (IIT Bombay), Dr.
    Shahid Malik (IIT Delhi) and Dr. Gaurav Saini (IIT Bombay)
Chapter 6: Sampling and Recovery of Signals with Spectral Sparsity
Dr. Shuchin Aeron (Tufts University) and Dr. Laxmeesha Somappa (IIT Bombay)
Chapter 7: Compressed Sensing
Dr. Shuchin Aeron (Tufts University) and Dr. Laxmeesha Somappa (IIT Bombay)

Contents

Contents

Chapter 1
Materials and Processing for Flexible Bioelectronics

1.1 Materials for Flexible Bioelectronics

A fully functional wearable device may integrate different physical, chemical and biological sensors with electronics for data collection, battery to power them and a display to show the results. In some cases, data can be wirelessly transmitted to a phone or a PC using flexible antennas and a wireless communication module, which can be used instead of a display. In some other bioelectronic applications, there may be actuators such as electrodes and circuits for electrical stimulation or drug delivery. A conceptual schematic layout of an example bioelectronic platform realized on a flexible substrate is shown in Fig. 1.1 [1].

Since bioelectronic devices are expected to interact intimately with the human body, their flexibility and biocompatibility with the underlying human tissue is of paramount concern. Mechanical flexibility (or stiffness) of the material is captured by its Youngs' modulus and varies widely based on the tissue or organ you are interfacing with as shown in Fig. 1.2 [2, 3]. As Fig. 1.2 suggests, there are large differences in the mechanical properties between different human tissues and materials (silicon, metal electrodes) used in conventional electronic devices. Notably silicon-based devices made using conventional photolithography are hard and brittle and not compatible with majority of tissue or organs such as brain or skin, and thus poorly suited for direct integration with them. There are two ways to minimize this gap in flexibility between and underlying material and the target tissue, one is to replace these electronic materials with more soft options. The first is a bottom-up approach where each material and device is flexible and stretchable at all length scales and under integration. The second approach is to miniaturize rigid devices and integrate them within a flexible polymeric material those that possess

Dr. Sameer Sonkusale (Tufts University) and Dr. Narendra Kumar (Boston College) contributed to this chapter.

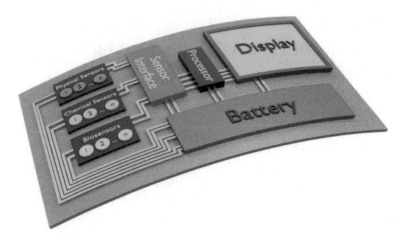

Fig. 1.1 A conceptual schematic of flexible bioelectronic device carrying different sensors, electronics to provide interface and data processing, a battery to power the electronics, and an electronic display for results. Some bioelectronic platforms may have an antenna and wireless communication device instead of a display, while some other platforms may also have actuators such as electrical stimulators or chemical drug delivery modules as part of the platform (not shown in the figure) [1]

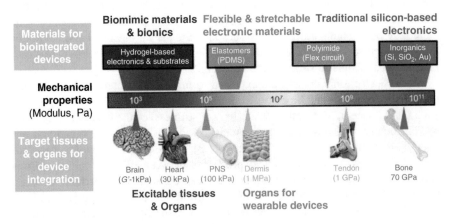

Fig. 1.2 Young's modulus of different target tissues and organs, compared to different materials to realize bioelectronic devices [3]

the elastic moduli close to the human tissues. Options to reduce thickness of bulk rigid materials such as thinning or making ribbons and using geometric patterning such as using serpentine layout imparts are also utilized to achieve flexibility. Going with the first approach requires extensive investigation on the properties and their implementation to develop electronic devices—several efforts are already being made since last 15 years with several successful examples [4–8]. The second approach is more convenient and practical, as one only needs to focus on developing smarter ways of integration that can result in a flexible bioelectronic device using rigid electronic and sensing components, or thinning bulk materials and patterning them [8–11].

We first begin our discussion by going through some choice of materials for substrates and materials for realizing electronics and sensors. Following this, we cover some standard methods of processing and fabrication of flexible devices and ways to integrate them. We also separately dedicate discussion on the second approach listed above to integrated miniature conventional electronic devices on flexible substrates to realize flexible and/or stretchable bioelectronics.

1.2 Substrates

Substrates serve as a foundation over which sensors, electronics, and other functional components are built to realize integrated devices. Therefore, their choice is crucial in determining the flexibility, biocompatibility, and performance of wearable flexible bioelectronic devices. The choice also depends on application. For example, one may need them to be transparent. In other application, they may need to be resistant to humidity and temperature, and while in some other cases, swelling of the substrates may be acceptable or even desirable. There are two major classes of materials being utilized as substrates, one is a class of polymeric materials and others are cellulosic substrates, which include papers and fibers/textiles.

1.2.1 Polymeric Materials

The most common substrates for flexible electronic devices are polymers such as polyethylene terephthalate (PET), polyethylene naphthalate (PEN), polyimide (PI), parylene, and polydimethylsiloxane (PDMS). PET substrates fulfill most of the requirements for substrates for flexible bioelectronics such as flexibility, chemical inertness, thermal and mechanical stability, and smoothness. Cost-effectiveness and extensive utility of these substrates in other areas such as for food packaging, liquid containers, and face shields make them highly desirable. PEN provides comparatively better thermal stability and chemical resistance compared to PET and is also a popular choice when conventional clean room fabrication process is desired to make devices. PI is another popular flexible material for the same reasons and is also heavily used in flexible printed circuit boards. Another material that is also FDA approved for medical and implantable devices is parylene, which is a highly inert substrate and is very popular as hermetic coating in implantable devices. They have been used for making electrode arrays for direct brain implants or making peripheral nerve interfaces. However, one of the main challenges with the materials listed above is that while they are flexible, their Young's modulus may still be quite higher than skin or brain, the organs they are used to interface with. Polydimethylsiloxane (PDMS) is an excellent choice for implantable sensors due to their high stretchability and Young's modulus closer to skin and other implantable tissues. They also show high biocompatibility and can be used for fixing sensors and devices directly on body parts especially at joints. They meet most of the criteria for future flexible bioelectronics because they are cost-effective, robust, and compatible to human tissues and conventional microfabrication techniques. It also offers opportunity to couple well-established microfluidics technology on PDMS whenever required for such devices and can be integrated together with electronics. PDMS is also a well-suited elastomer to be utilized as a matrix to integrate different electronic components to realize a wearable device, more details are provided in the later part of this chapter. However, PDMS is not always chemically inert. It also has its own limitations such as swelling in organic solvents and nonspecific adsorption of cells and proteins in case of microfluidics and implantable applications.

Hydrogels (e.g., poly(ethylene glycol), poly(vinyl alcohol), collagen, gelatin, chitosan, alginate) are class of materials which have made significant advancement to provide flexible interconnects and conducting electrodes [12–15]. They possess strong stretchability, biocompatibility, and closely match elastic moduli to human tissues. Moreover, they provide flexibility of modulating their electrical, mechanical, and biological properties making them a unique bridging material to biological world [14]. Hydrogels are made from cross-linked polymeric network and have high degree of hydrophilicity because of excess amount water contents, which easily allow the transport of biological and chemical analytes desired for bioelectronic devices [16]. All these properties of hydrogels rendered their growing utilization in different bioelectronic devices and being utilized either as interface material or to form composite matrix to modulate the properties of metallic and carbon-based materials [13]. Recent advancements in the field are utilizing hydrogels to develop tissue like wearable bioelectronic platforms [17, 18].

1.2.2   Fibers/Textiles

Substrates based on fibers, threads, and textiles offer a highly versatile option to realize flexible bioelectronics [19, 20]. First of all, they are cheaper than other substrates, and benefit greatly from established textile making and processing infrastructure. Fibers and threads also offer a rich diversity of material options (e.g., rubber, nylon, cotton) which can be knitted, woven, or patterned to achieve flexibility or stiffness over a wide range of values for Young's modulus. Tunable mechanical properties and long lifetime of fibers and textiles provide a promising platform to realize wearable sensing devices. Compared to top-down fabrication approaches to realize electronic devices on other substrates, textile-based flexible bioelectronics utilize more bottom-up approach. This is because, textile-based substrates are nonuniform and porous making uniform patterning and coating harder. Better results can be achieved if functional smart threads are first made using conventional dyeing or coating approach, and then threads are sewn or knitted together to make functional devices [19, 20]. Several well-known knitting and weaving processes allow integration of the functional fiber-based electrodes and electronic components on desired textiles. The major advantage of fabrics/textile is that they do not need cleanroom-based fabrication methods and use ambient processing. A sister substrate to textile, and one that is just as ubiquitous is paper, a cellulosic substrate that is environmentally friendly and amenable to simple processing to make functional devices such as inkjet printing or screen printing. However, the paper-based substrates are suitable for short-term temporary applications as they are biodegradable and have poor thermal and chemical stability.

In general, the recommended substrates for flexible bioelectronics depend on the application and the type of functional devices one intends to make on them. As mentioned, the choices can range from PET, PEN, and PI for long-term flexible devices, to PDMS as matrix to fabricate and integrate conventional rigid electronic devices and circuits to textile-based substrates for wearable applications. We should mention here, as we did earlier, that in near term, the approach of integrating conventional electronics over flexible elastomeric support seems more reasonable over approaches that rely on fully novel material systems.

1.3 Electronic Functional Materials

As shown in Fig. 1.1, flexible bioelectronic platform consists of sensors, transistors, and other functional components. Herein we summarize different choices of materials used in making these components. The list may not be complete but provides a quick insight into available options.

1.3.1 Organic Materials

Organic electronic materials (OEMs) have been widely studied and utilized to develop flexible electronic devices such as solar cells, organic LEDs, and batteries [21–23]. OEMs are also interesting active materials for flexible bioelectronics because of their synthetic tunability, biocompatibility, and mechanical flexibility, which make them suitable to integrate with cells and biological tissue [24, 25]. Major OEMs being utilized as active semiconducting materials in bioelectronics are poly(3,4-ethylenedioxythiophene):poly(styrene sulfonate) (PEDOT:PSS), poly (3-hexylthiophene-2,5-diyl) (P3HT), and phenyl-C_{61}-butyric acid methyl ester (PCBM) [26, 27]. PEDOT:PSS is a class of highly conducting OEMs consisting of a polymer mixture of positively charged PEDOT with negatively charged PSS, and their conductivity can be tailored by choosing the ratio of these two compounds and their resulting particles size in dispersion. It is commonly used as a semiconductor in a transistor operating in depletion mode [28, 29]. P3HT is a p-type organic semiconductor that works in accumulation mode when utilized as an active channel material in transistors because of its high mobility >0.1 cm^2/V.s and low operating voltage as compared to PEDOT:PSS [30]. Both these OEMs are also being employed to fabricate organic electrochemical transistors for developing flexible biosensing devices [31, 32]. Doping or de-doping of electrolytes in PEDOT:PSS and P3HT can be achieved using a gate electrode in an aqueous environment providing transistor functionality [33, 34]. OEMs allow exchange of ions between electrolyte/biofluids and the organic semiconductor resulting in the involvement of the whole volume of the film interacts with biological environment, not just its surface, a feature which can be exploited to develop powerful biosensors and bioactuators [24].

1.3.2   Metal Oxides

Metal oxides and their nanostructures are another class of materials those can be utilized for flexible devices [35]. These materials are available in the form of dielectrics, semiconductors, and conductors. Zinc oxide (ZnO), indium oxide (In2O3), indium gallium zinc oxide (IGZO), indium tin oxide (InSnO), and many more combination possesses semiconducting properties and are being utilized as active layer in fabricating transistors [36]. Metal oxide semiconductors possesses excellent electrical performance (mobility >10 cm^2 V^{-1} s^{-1} at room temperature processing) along with their compatibility of fabrication over flexible substrates. On the other hand, aluminum oxide (Al_2O_3), titanium oxide (TiO_2), yttrium oxide (Y_2O_3), zirconium oxide (ZrO_2), tantalum oxide (Ta_2O_5) are utilized as gate dielectrics because of their high dielectric constant and smoother interface with metal oxide semiconductors [36]. For example, IGZO thin-film transistor showed a mobility of 14.5 cm^2 V^{-1} S^{-1} processed below 350 °C, while it showed a mobility of 7.5 cm^2 V^{-1} S^{-1} processed at room temperature (RT) [37, 38]. These materials can be processed either by physical deposition systems at low temperature or through solution processing techniques discussed earlier. The solution processing of these materials by first preparing the nanomaterial precursor inks using sol-gel process offers a low-cost route that can then be used to make devices using spin/dip coating and screen/inkjet printing [39]. However, most of these materials still require high temperature (≥300 °C) annealing and suffer from limited flexibility and stretchability; additional modifications in the supporting substrates and encapsulation are needed to overcome these challenges [39–41].

1.3.3   Carbon-based Materials

Carbon-based nanomaterials such as graphite, carbon nanotubes (CNTs), and graphene are expected to provide a more natural and biocompatible interface with the living universe that is entirely based on carbon. CNTs and graphene have shown great promise in the field of flexible bioelectronics with several examples from neural electrode arrays to flexible transistors or electronic tattoos [42–45]. Specifically, the CNTs can be metallic or semiconducting, grown by chemical vapor deposition (CVD) in bulk and easily dissolved in solvents to make ink formulations that can be used to fabricate electrodes/transistors using spin/spray coating, and inkjet/screen printing [45–48]. For instance, high-performance and low-voltage (3 V) flexible analog and digital circuits were fabricated using CNTs in solution form. The transistors fabricated on 4 inch polyimide substrate showed high mobility values ~34 cm^2/V.s with 5 mm bending radius [49]. There is well established growth and transfer process of graphene making it one of the popular sensing and electronic materials at present. Recently, the roll-to-roll fabrication of these 1D/2D material-based devices has been reported indicating their potential for future flexible electronics [50, 51]. For electrochemical sensing, it is well known that graphitic carbon provides a stable working electrode. Carbon nanomaterial-based inks and paste are readily available commercially for screen printing of working electrodes and counter electrodes on any flexible substrates to make disposable electrochemical sensors [52–56]. Furthermore, the stability of carbon-based electrodes and materials in physiological environment, flexibility, and biocompatibility make it an excellent choice for future flexible bioelectronic devices [45].

1.3.4 Conductive Inks and Liquid Metals

Conductive inks (gold, silver, carbon, etc.) are easily available in the market and can be printed/coated/pasted on any kind of substrate using drop casting or inkjet printing. There are pens available filled with these formulated inks and are extremely helpful to draw the desired connection to make the circuitry. The properties of these inks can also be modulated as per the requirement of the application. There is also a eutectic alloy of gallium–indium that is metallic liquid at room temperature and has interesting wettability on select surfaces; this can be used to realize flexible interconnects that are self-healing in presence of breakage [57, 58].

1.4 Materials Processing

How do we go from materials to functional devices such as sensors or electronics is the subject of this discussion on materials processing. Since we are interested in making devices on flexible polymeric substrates, it is important that processing is done at low temperature since these substrates cannot handle high temperature. Solution-based methods and bottom-up approaches are more suitable. We cover two different techniques for materials processing to make flexible devices, one is a bottom-up approach using different functional inks and patterning them directly on flexible substrate, while the other is based on transfer printing from one substrate to another.

1.4.1 Printing from Inks

The various solution processing techniques such as spin, dip, spray, blade coating, and electrochemical deposition can be performed at low temperatures followed by evaporating the solvents at temperature susceptible to flexible substrates. In these processes (e.g., spin coating), the desired materials are synthesized as nanoparticles in a precursor having different solvents. After that, the solution is filtered to remove any remaining clusters. Coating is done by pouring the solution over the surface of substrates and running the spin coater at optimized spinning speed. In this process, centrifugal force spins out the solution from center to outward direction which depends on the speed chosen to obtain the desired thickness and uniformity. In the dip coating, the substrates are dipped in and out of the prepared solution at different speeds with varying number of dipping cycles. Spray coating, the substrate rotated at fixed speed and solution sprayed though nozzles, spray timing, and speed of substrate rotation decide the thickness and uniformity of film. The blade coating is done by injecting the ink (solution) while maintaining a fixed distance between knife/blade and moving substrates. The optimized annealing is done to evaporate the solvents followed by characterization of materials properties. See Fig. 1.3 for conceptual representation of these methods [59]. While these methods provide a uniform coat needed for any material, one needs to employ other methods to patterning them.

In conventional lithography, optical masks and photoresists are used to identify different areas for either ion implantation, or metal sputtering or oxide growth. However, this is not practical for realizing devices on flexible substrates due to their sensitivity to high temperature. Instead, solution-based patterning of different materials will enable realization of multilayered devices such as transistors or sensors. Some of the available methods for patterning with different inks are shown

Fig. 1.3 Examples of solution-based coating process that allows low temperature material deposition on any flexible substrate [59]

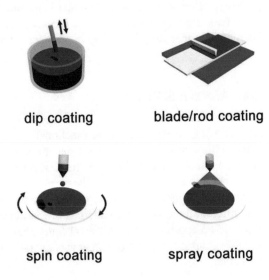

dip coating   blade/rod coating

spin coating   spray coating

Fig. 1.4 Some methods for
patterning different materials
on flexible substrates [59]

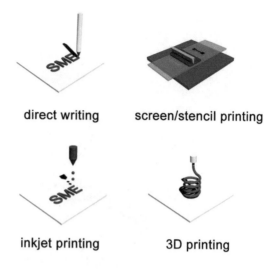

direct writing   screen/stencil printing

inkjet printing   3D printing

in Fig. 1.4. In case of transistors, patterning will help defining areas of source and drain regions where metal is deposited, and areas for semiconductor deposition.

Some methods for patterning different materials on flexible substrates are shown in Fig. 1.4. Direct writing uses a writing pen or pencil containing target material for direct writing on flexible substrates. The wettability of ink on substrate is key for realizing devices using this method. One popular example is the use of pencil on paper approaches to realize sensors and interconnects on paper substrates [60]. This approach is however slow and is not high throughput and therefore not scalable. However, it is an excellent approach to realize proof-of concept devices in a laboratory setting.

Printing offers a more scalable route for fabrication of flexible electronic devices over a large area. One such high-throughput method to realize large area sensors and electronics on flexible substrates is screen printing. In this approach, a hard mask made of metal or polymer is used. Exposed areas in the mask define areas where deposition is to take place. An ink is applied using a squeegee which transfers the ink through the hard mask on to the underlying flexible substrate. The mask is then removed leaving behind patterned areas with ink. Repeated application of same and different inks through different masks can create complex multilayered structure such as transistors. For example, one can use one mask to screen print metal source and drain areas, followed by another mask for deposition of semiconducting inks. Yet another mask is needed for both insulating gate and metal gate inks [61]. The resolution of the devices depends on the resolution of the mask and the wettability of ink on the substrate. Moreover, for multilayered stacked devices, registration and alignment errors limit over all resolution. Tens to hundreds of microns of resolution can be easily achieved using screen printing.

Several other printing techniques such as inkjet printing, 3D, roll-to-roll processing have been proposed to scale up fabrication of flexible bioelectronic devices [62, 63]. In particular, commercial inkjet printing used for printing color inks on

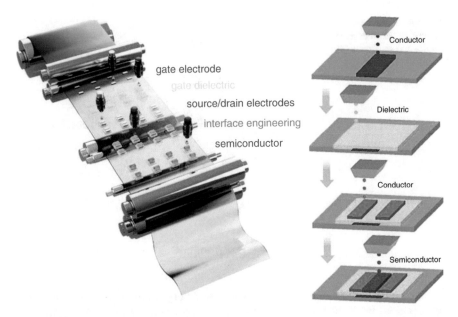

Fig. 1.5 Roll-to-roll sequential inkjet printing of devices on flexible substrate (left). A back-gated transistor fabricated using inkjet printing involves use of multiple inks to realize multilayered transistor stack [63, 64]

paper can be leveraged for making functional smart devices on flexible substrates by changing the ink formulation. Inkjet printing can also be scaled up for batch processing in a roll-to-roll format with multiple nozzles for delivery of different inks. See Fig. 1.5 for a conceptual representation [63]. An example of an individual transistor in this roll-to-roll printing workflow is also shown in Fig. 1.5. Resolution and accuracy of inkjet printing has been improving considerably and depends on the precision of droplet generation at the nozzle and the wettability of the ink on the substrate. One can achieve 2 ~ 5 micron level resolutions easily using inkjet printing. Moreover, inkjet printing has higher registration accuracy, which allows the fabrication of devices with complex stacked structures without masks This is considerably better than screen printing-based method discussed earlier.

Beyond nozzle-based inkjet printing and screen printing, one can also utilize offset, flexography, and gravure-based printing. Specifically, gravure printing allows high-resolution and high-throughput fabrication. However, it suffers from poor registration accuracy for multilayered devices and use of high viscosity inks generates residues [65]. The noncontact nature of inkjet printing is still preferred since it can realize high-quality printed features without the need for rinsing.

1.4.2 Transfer Printing

Some of the high-performance materials may not be directly processable at low temperatures. A hybrid approach can then be used for realizing flexible electronics by fabricating rigid devices using conventional high yield bulk processing followed by thinning or other post-processing methods and transferring them to flexible substrates. Transfer printing technique is an emerging area which facilitates the fabrication of mico-/nano-scale devices over flexible or any other desired substrates. As shown in Fig. 1.6, in this process, the structures/devices (inks) are fabricated over rigid substrates (donor) utilizing sacrificial layers. The retrieval of inks from the substrates is done utilizing stamp followed by their delivery over desired flexible (receiver) substrates [66, 67]. Crucial factors in this technique are the sacrificial layer, ink/stamp adhesion, and receiver/ink adhesion; these play critical role for the successful transfer of the structures over flexible substrates and needed to be optimized [68, 69].

Similarly, one can also explore transfer of some novel 2D materials. CVD graphene is an appropriate example of such material which can easily be transferred over any flexible substrates using standard wet (chemical etching method) and dry

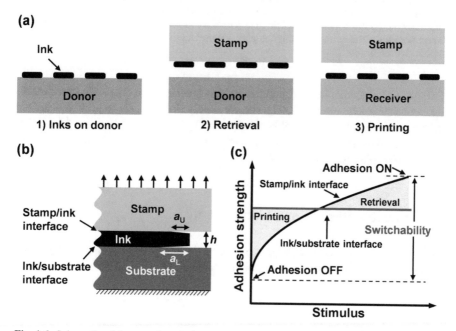

Fig. 1.6 Schematic of the transfer printing process. (1) Inks are prepared on the donor substrate in a releasable manner. (2) Retrieval process: an elastomer stamp is used to retrieve the inks. (3) Printing process: inks are printed onto the receiver substrate. The two interfaces in the stamp/ink/substrate structure. Adhesion strength modulated by external stimulus, showing the high (ON) and low (OFF) adhesion state and the switchability [66, 67]

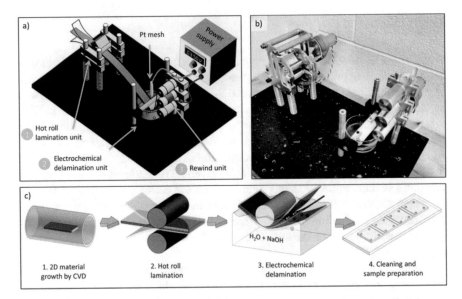

Fig. 1.7 Schematic of the R2R transfer setup consisting of a hot roll lamination unit, electrochemical delamination unit, and rewind unit. (**b**) Actual implementation of the transfer setup. (**c**) Transfer process flow starting with growing the 2D material on a metal film, laminating it in between plastic substrates on top and bottom by applying pressure and heat, electrochemically separating the 2D layers from the metal surface, rinsing the plastic substrates, and gluing them on glass slides for further characterization [50]

[70] (using laminator) transfer process and has been extensively studied for flexible bioelectronic applications [42, 71]. Efforts are also being made to grow the desired 2D materials directly over flexible substrates [72]. To achieve large area fabrication of 2D material-based devices over flexible substrates, roll-to-roll transfer process can be used (see Fig. 1.7) [50, 73].

1.5 Flexible Devices and Components

In this section, we will look at few select examples of different devices and components, primarily electronic devices fabricated on flexible substrates.

1.5.1 Electrodes

Electrodes are key functional unit in realization of sensors, capacitors, and batteries. They are also used in realization of drain, source, and gate terminals in transistors.

Three-electrode configuration is commonly employed to realize electrochemical sensors. In electrochemical sensors, a working electrode is made from carbon or gold and functionalized with nanoparticles, antibodies, or aptamers or enzymes with and without redox mediators. A counter electrode made from carbon or platinum and reference electrode made from Ag/AgCl or Pt is used. This three-electrode geometry is used for potentiometric and amperometric detection. Electrodes are also used in interdigitated configuration to realize impedimetric sensors; the details about these systems will be discussed in the next chapter. Fabrication of metal electrodes can be done using one of the several methods discussed in the previous section. As an example, screen printing or inkjet printing has been a popular choice to fabricate these electrodes over a variety of flexible substrates. Ink formulations range from carbon-based materials, conducting polymers, and their respective nanostructures to realize these electrodes.

Miniaturized flexible energy storage devices (batteries or supercapacitors) are highly desired for the operation of flexible bioelectronic devices. Supercapacitors have drawn a lot of attention as they offer high-power, long life cycle, and environment-friendly nature and bridging the gap between batteries and classical electric capacitors [74]. Specifically, printed micro-supercapacitors have become an excellent choice for flexible bioelectronics as they provide a whole range of simple, low-cost, time-saving, versatile, and environmental-friendly manufacturing technologies [75]. Recently, an integrated flexible micro-supercapacitor device with wireless charging was realized, which opens the path for an inbuilt charging system for wearable bioelectronics [76].

1.5.2 Transistors

Field-effect transistors have been extensively studied for flexible bioelectronics to interface with flexible sensors and actuators. A typical transistor consists of source/drain electrodes, a semiconductor, a gate dielectric, and a gate electrode. Since conventional silicon-based transistors are not compatible for fabrication over flexible substrates due to high thermal budget, thin-film-based field-effect transistors are being employed for flexible bioelectronics. Thin-film semiconductors being used in such devices are oxides, organic, and nanowire, specifically carbon nanotubes [49]. Also except pH sensing, these semiconducting thin-film transistors operated in liquid gating (without dielectric) mode by utilizing the concept of electric double layer as thin gate dielectric forms at semiconductor/liquid interface (see Fig. 1.8) [77].

Organic electrochemical transistor (OECT) fabricated over flexible substrates and fabric provided significant advancement to the field [31, 32]. In organic electrochemical transistor, electrochemical doping or de-doping of an organic

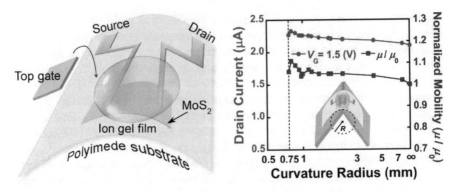

Fig. 1.8 A schematic of ion gel-gated MoS2 transistor over polyimide substrate (left), variation in drain current, and mobility upon bending at different curvature radius [77]

semiconducting channel is employed for transistor action. Figure 1.9 shows a typical example of OECT fabricated directly over nylon fiber. First, the source/drain contacts of the transistor were fabricated by depositing Cr/Au (10/100 nm) on the fiber and protecting the channel region (0.4 mm) in the middle. Then, the PEDOT: PSS, a popular choice for semiconducting channel in OECT, was deposited over the fiber except the contact area at both ends followed by a parylene layer to passivate the electrodes to eliminate faradic current during operation. Another piece of fiber was deposited with Ti/Pt to form a gate electrode which was further passivated with parylene to define geometric gate area in the middle and leaving open metal part at the ends to supply the voltage. Then both wires are placed together and exposed together with a drop of liquid electrolyte solution. The transfer characteristics of the transistor was obtained by measuring drain current at varying voltage applied through gate electrode (Fig. 1.9e). Crucial parameters to be taken care to develop flexible transistors are the conductivity of the electrodes at different bending angles, which depends on the uniformity of the coated metal and semiconductors as well as their adhesion to the chosen flexible substrates.

Figure 1.10 shows flexible IGZO TFT array fabricated over PET substrate at RT [37]. AlOx as dielectric as well as passivation layer was obtained by anodization of Al at RT. A liquid-gated IGZO TFT has recently been reported as a flexible biosensor, suggesting its future application in bioelectronic devices [78].

Graphene-based transistors have shown great potential because of their high sensitivity in detection of various disease biomarkers, scalability, biocompatibility, and ease of incorporation on conventional and flexible substrates [79–82]. Recently, some graphene field effect transistor (GFET)-based biosensors have been reported over flexible substrates that include polyimide, PET, and paper networks [83–85]. Figure 1.11 shows one of the examples of fabricating flexible GFET on wafer scale [86]. In this process, first the CVD graphene was coated with parylene followed by

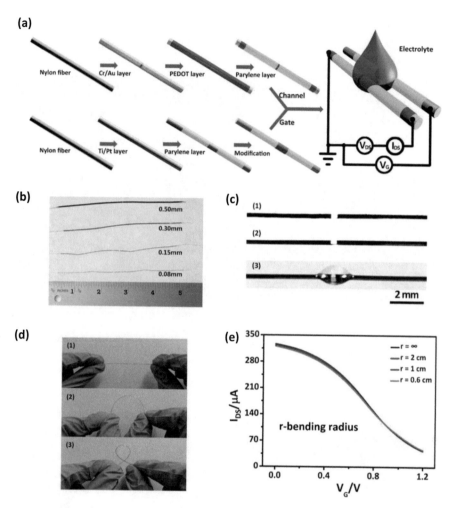

Fig. 1.9 (**a**) Fabrication process of an OECT device over nylon fiber, adopted design has one fiber that contains source drain electrode and semiconductor while another wire works as a side gate electrode when placed alongside. (**b**), (**c**) Images of fabricated devices with different diameter of fibers. (**d**, **e**) Transistor characteristics at different bending radius [32]

etching growth substrate (Cu) and transferred over SiO2/Si. Then whole fabrication was done over SiO2/Si substrate which includes source drain contacts, parylene coating as dielectric and gate electrode. Then finally transistors were peeled off keeping parylene as flexible supporting substrate.

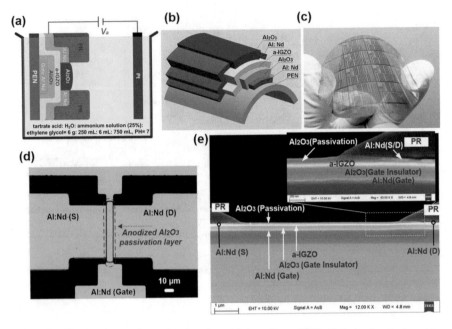

Fig. 1.10 An a-IGZO-based thin-film transistor fabricated over PEN substrate at room temperature by utilized anodized Al_2O_3 as gate dielectric [37]

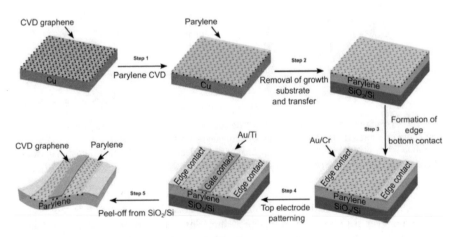

Fig. 1.11 A graphene transistor fabricated over parylene substrate, the fabrication was done over rigid SiO2/Si substrate having parylene as intermediate layer and whole transistor was peeled off post-fabrication [86]

1.6 Packaging and Integration

Once devices such as sensors or transistors are fabricated, they need to be integrated together into a system with interconnects, etc. In fact, the individual sensors or transistors may not even be flexible, and the overall flexibility is achieved through packaging and integration of these discrete components onto a flexible polymeric platform.

1.6.1 Methods

There are two different ways to achieve integration and packaging of devices into system. One method is the co-fabrication of the active elements directly over flexible platform along with the interconnects using one of the several methods for fabrication discussed earlier. In other cases, devices are fabricated first using conventional processing and then post-processed such as thinning to achieve flexibility and transferred to flexible substrates where interconnects are then added. A more direct and near-term route to realizing flexible bioelectronic devices is to start with rigid functional devices and transferring them on flexible substrates. Such an approach takes advantage of established cleanroom processing of conventional devices such as silicon Complementary Metal Oxide Semiconductor (CMOS) Integrated Circuit (IC) to realize reliable and complex functionality. The focus is then on integration and packaging of these rigid miniaturized devices on flexible substrates. In essence, what we have achieved is a flexible (and possibly stretchable) printed circuit board with rigid micro-/nanoelectronics chip. Achieving flexibility using metal interconnects in all these approaches requires using geometrical designs with different architectures to achieve flexibility. For example, wavy serpentine structure has shown great promise to obtain stretchability without sacrificing the conductivity. Figures 1.12 and 1.13 provide the examples of the realization of enhanced stretchability more than 800% using lateral spring design with copper interconnects, and demonstration of a smart bandage for monitoring chronic wound consisting of integrated pH and temperature sensors made with serpentine gold interconnects [87, 88]. This is possible because wavy structures are sustainable under strains due to their properties of varying wavelengths and wave amplitudes [8]. Such serpentine interconnects made from gold or other metal that are otherwise not stretchable in bulk form and can now be used to realize flexible and stretchable interconnects over a large strain >100%. The serpentine interconnects possess lower elastic moduli and high bending radius with minimum areal density. These excellent properties of serpentine interconnects provided excellent choice to develop epidermal bioelectronic devices [89, 90].

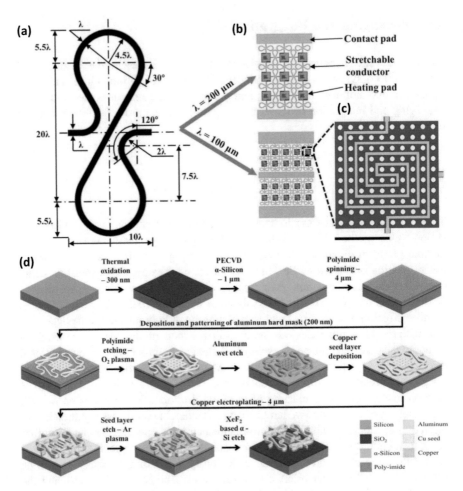

Fig. 1.12 A CMOS-compatible process to fabricate stretchable thermal patch using copper interconnects. Design of the stretchable lateral spring provides up to 800% uniaxial stretchability [87]

1.6.2 Package Assembly on Silicon CMOS Platform

As mentioned earlier in this book, a more practical route to flexible bioelectronics is to utilize CMOS silicon IC for electronic functionality to interface with sensors and electrodes. This allows us to leverage advanced computing and signal processing power available in silicon CMOS process with large area flexible sensors and bioelectronics. In this section, we focus on packaging and integration of such CMOS platforms with the goal of achieving flexible bioelectronic systems.

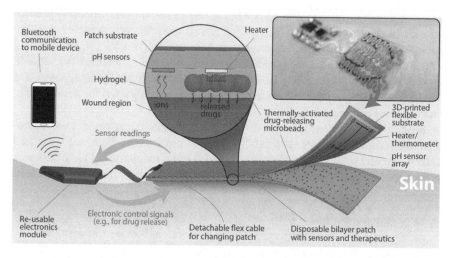

Fig. 1.13 Conceptual illustration of a smart bandage made using serpentine metal (gold) interconnects containing an array of flexible pH sensors, heater [88]

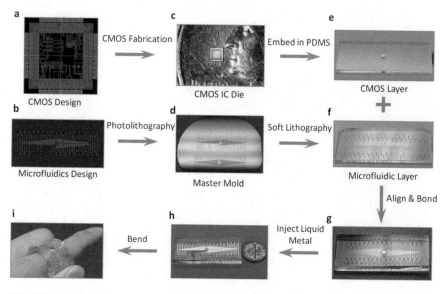

Fig. 1.14 Fabrication process to integrate conventional silicon-based CMOS chips over PDMS substrates. Process utilizes prepatterned microfluidic channels to inject liquid metal to form interconnects [9]

Figure 1.14 shows a successful demonstration of the solid integrated circuit (IC) chip over flexible PDMS platform having interconnects made by injecting liquid metals through microfluidics [9]. Liquid metal interconnects provide high stretchability and flexibility without losing their conductive properties and making them an excellent choice for flexible bioelectronics [10]. Compared to conventional metal electrodes which need to be laid out in serpentine or other geometric patterns to withstand stress from stretching, liquid metals are inherently stretchable due to their liquid constitution. Such interconnects take the form of a microfluidic channel and can easily be 3D printed as desired for specific application/platform [91]. PDMS also provides a biocompatible, stable, chemically/physiologically susceptible, and flexible packaging support which has been most commonly used for microfluidics in healthcare and other industries.

Another approach, as alluded to earlier, uses conventional metals in wavy serpentine like format to support stretchability without breakage or fissures. Figure 1.15 shows a typical fully integrated sensor for daily stress monitoring [11]. This device comprises in a pair of skin-conformal thin-film sensors and stretchable membrane

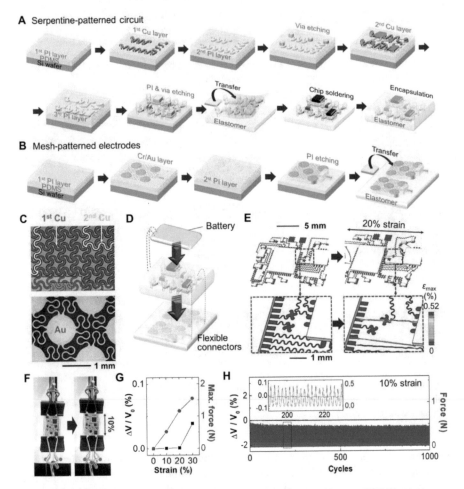

Fig. 1.15 Fabrication of serpentine patterned circuit and mesh electrodes over PDMS as interconnects to integrate and encapsulate chips and battery [11]

wireless circuit, fabricated using multilayered nanostructured materials and integrated over a soft elastomeric membrane. The serpentine patterned interconnects were utilized to enhance the flexibility and stretchability of these devices. First, the serpentine circuit and mesh of electrodes were fabricated over silicon and then integrated together over elastomer support using transfer printing. After that, all the required electronic components were also integrated at desired locations. As shown in Fig. 1.15g, sensor showed negligible variation in the electrical performance up to 20% of tensile strain, and no significant effect of cyclic loading until 1000 cycles with 10% tensile strain (see Fig. 1.15h). Inset shows the magnified view of signal changes during the cyclic loading. Another method for making stretchable electronics is through using 3D integrated multilayer approach built using layer by layer with the help of transfer printing of fabricated stretchable circuits over each layer of elastomer [92].

The fabrication of sensing devices and interconnects on directly over flexible support has been extensively investigated and realized for flexible bioelectronics. Nanomaterials such as metallic CNTs, graphene inks, and silver nanoparticles have been widely used to develop wearable and stretchable devices [93]. These nanomaterials can act as a sensing material as well as can form interconnects for circuitry; they can also be assembled directly in to elastomeric matrices [94]. Such a combination of materials takes care of both sensing, integration, and packaging issues for flexible bioelectronics.

References

1. F. Liu, A.S. Dahiya, R. Dahiya, A flexible chip with embedded intelligence. Nat. Electron. 3(7), 358–359 (2020)
2. C. Xu, Y. Yang, W. Gao, Skin-interfaced sensors in digital medicine: From materials to applications. Matter 2(6), 1414–1445 (2020)
3. C.J. Bettinger, Recent advances in materials and flexible electronics for peripheral nerve interfaces. Bioelectron. Med. 4(1), 1–10 (2018)
4. W. Gao et al., Flexible electronics toward wearable sensing. Acc. Chem. Res. 52(3), 523–533 (2019)
5. Y. Liu, M. Pharr, G.A. Salvatore, Lab-on-skin: A review of flexible and stretchable electronics for wearable health monitoring. ACS Nano 11(10), 9614–9635 (2017)
6. J. Kang, J.B.-H. Tok, Z. Bao, Self-healing soft electronics. Nat. Electron. 2(4), 144–150 (2019)
7. W.S. Wong, A. Salleo, Flexible electronics: materials and applications, vol 11 (Springer Science & Business Media, New York, 2009)
8. J.A. Rogers, R. Ghaffari, D.-H. Kim, Stretchable bioelectronics for medical devices and systems (Springer, Cham, 2016)
9. B. Zhang et al., Flexible packaging of solid-state integrated circuit chips with elastomeric microfluidics. Sci. Rep. 3(1), 1098 (2013)
10. M. Li et al., Liquid metal-based electrical interconnects and interfaces with excellent stability and reliability for flexible electronics. Nanoscale 11(12), 5441–5449 (2019)
11. H. Kim et al., Fully integrated, stretchable, wireless skin-conformal bioelectronics for continuous stress monitoring in daily life. Adv. Sci. 7(15), 2000810 (2020)
12. Das, S., et al., Processable, Ion-conducting hydrogel for flexible electronic devices with self-healing capability. Macromolecules, 53:11130 2020

13. H. Yuk, B. Lu, X. Zhao, Hydrogel bioelectronics. Chem. Soc. Rev. **48**(6), 1642–1667 (2019)
14. Y.S. Zhang, A. Khademhosseini, Advances in engineering hydrogels. Science **356**, 6337 (2017)
15. Z. Li et al., Gelatin Methacryloyl-based tactile sensors for medical wearables. Adv. Funct. Mater. **30**(49), 2003601 (2020)
16. F. Fu et al., Functional conductive hydrogels for bioelectronics. ACS Mater. Lett. **2**(10), 1287–1301 (2020)
17. C. Lim et al., Tissue-like skin-device interface for wearable bioelectronics by using ultrasoft, mass-permeable, and low-impedance hydrogels. Sci. Adv. **7**(19), eabd3716 (2021)
18. Y. Liu et al., Ultrastretchable and wireless bioelectronics based on all-hydrogel microfluidics. Adv. Mater. **31**(39), 1902783 (2019)
19. P. Mostafalu et al., A toolkit of thread-based microfluidics, sensors, and electronics for 3D tissue embedding for medical diagnostics. Microsyst. Nanoeng. **2**(1), 1–10 (2016)
20. J. Xia et al., Thread-based wearable devices. MRS Bull. **46**, 502–511 (2021)
21. M.T. Dang, L. Hirsch, G. Wantz, *P3HT: PCBM, best seller in polymer photovoltaic research*. (Wiley Online Library, 2011)
22. T.W. Kelley et al., Recent progress in organic electronics: Materials, devices, and processes. Chem. Mater. **16**(23), 4413–4422 (2004)
23. M. Berggren, D. Nilsson, N.D. Robinson, Organic materials for printed electronics. Nat. Mater. **6**(1), 3–5 (2007)
24. J. Rivnay, R.I.M. Owens, G.G. Malliaras, The rise of organic bioelectronics. Chem. Mater. **26**(1), 679–685 (2014)
25. M. Berggren, A. Richter-Dahlfors, Organic bioelectronics. Adv. Mater. **19**(20), 3201–3213 (2007)
26. W. Gao et al., Fully integrated wearable sensor arrays for multiplexed in situ perspiration analysis. Nature **529**(7587), 509–514 (2016)
27. Y. van de Burgt et al., A non-volatile organic electrochemical device as a low-voltage artificial synapse for neuromorphic computing. Nat. Mater. **16**(4), 414–418 (2017)
28. D.J. Lipomi et al., Electronic properties of transparent conductive films of PEDOT:PSS on stretchable substrates. Chem. Mater. **24**(2), 373–382 (2012)
29. M. Vosgueritchian, D.J. Lipomi, Z. Bao, Highly conductive and transparent PEDOT: PSS films with a fluorosurfactant for stretchable and flexible transparent electrodes. Adv. Funct. Mater. **22**(2), 421–428 (2012)
30. J.-F. Chang et al., Enhanced mobility of poly (3-hexylthiophene) transistors by spin-coating from high-boiling-point solvents. Chem. Mater. **16**(23), 4772–4776 (2004)
31. C. Liao et al., Flexible organic electrochemical transistors for highly selective enzyme biosensors and used for saliva testing. Adv. Mater. **27**(4), 676–681 (2015)
32. A. Yang et al., Fabric organic electrochemical transistors for biosensors. Adv. Mater. **30**(23), 1800051 (2018)
33. A. Giovannitti et al., Controlling the mode of operation of organic transistors through side-chain engineering. Proc. Natl. Acad. Sci. **113**(43), 12017–12022 (2016)
34. L.Q. Flagg et al., Anion-dependent doping and charge transport in organic electrochemical transistors. Chem. Mater. **30**(15), 5380–5389 (2018)
35. L. Petti et al., Metal oxide semiconductor thin-film transistors for flexible electronics. Appl. Phys. Rev. **3**(2), 021303 (2016)
36. W. Xu et al., Recent advances of solution-processed metal oxide thin-film transistors. ACS Appl. Mater. Interfaces **10**(31), 25878–25901 (2018)
37. X. Xiao et al., Room-temperature-processed flexible amorphous InGaZnO thin film transistor. ACS Appl. Mater. Interfaces **10**(31), 25850–25857 (2018)
38. N. Kumar et al., Interface mechanisms involved in a-IGZO based dual gate ISFET pH sensor using Al2O3 as the top gate dielectric. Mater. Sci. Semicond. Process. **119**, 105239 (2020)
39. J.W. Park, B.H. Kang, H.J. Kim, A review of low-temperature solution-processed metal oxide thin-film transistors for flexible electronics. Adv. Funct. Mater. **30**(20), 1904632 (2020)

40. X. Li, J. Jang, Stretchable oxide TFT for wearable electronics. Inf. Display **33**(4), 12–39 (2017)
41. Y.-H. Kim et al., Highly robust neutral plane oxide TFTs withstanding 0.25 mm bending radius for stretchable electronics. Sci. Rep. **6**(1), 25734 (2016)
42. S. Kabiri Ameri et al., Graphene electronic tattoo sensors. ACS Nano **11**(8), 7634–7641 (2017)
43. K. Wang et al., Neural stimulation with a carbon nanotube microelectrode Array. Nano Lett. **6**(9), 2043–2048 (2006)
44. D.-W. Park et al., Graphene-based carbon-layered electrode array technology for neural imaging and optogenetic applications. Nat. Commun. **5**(1), 1–11 (2014)
45. D.J. Lipomi et al., Skin-like pressure and strain sensors based on transparent elastic films of carbon nanotubes. Nat. Nanotechnol. **6**(12), 788–792 (2011)
46. A. Chortos et al., Mechanically durable and highly stretchable transistors employing carbon nanotube semiconductor and electrodes. Adv. Mater. **28**(22), 4441–4448 (2016)
47. F. Molina-Lopez et al., Inkjet-printed stretchable and low voltage synaptic transistor array. Nat. Commun. **10**(1), 2676 (2019)
48. X. Cao et al., Fully screen-printed, large-area, and flexible active-matrix electrochromic displays using carbon nanotube thin-film transistors. ACS Nano **10**(11), 9816–9822 (2016)
49. T. Lei et al., Low-voltage high-performance flexible digital and analog circuits based on ultrahigh-purity semiconducting carbon nanotubes. Nat. Commun. **10**(1), 2161 (2019)
50. M. Hempel et al., Repeated roll-to-roll transfer of two-dimensional materials by electrochemical delamination. Nanoscale **10**(12), 5522–5531 (2018)
51. S. Lu et al., Flexible, print-in-place 1D–2D thin-film transistors using aerosol jet printing. ACS Nano **13**(10), 11263–11272 (2019)
52. J. Wang, Carbon-nanotube based electrochemical biosensors: A review. Electroanalysis Int. J. Devoted Fundam. Pract. Asp. Electroanalysis **17**(1), 7–14 (2005)
53. J. Schnitker et al., Rapid prototyping of ultralow-cost, inkjet-printed carbon microelectrodes for flexible bioelectronic devices. Adv. Biosyst. **2**(3), 1700136 (2018)
54. W. Jia et al., Electrochemical tattoo biosensors for real-time noninvasive lactate monitoring in human perspiration. Anal. Chem. **85**(14), 6553–6560 (2013)
55. B. Ciui et al., Wearable wireless tyrosinase bandage and microneedle sensors: Toward melanoma screening. Adv. Healthc. Mater. **7**(7), 1701264 (2018)
56. J. Min et al., Wearable electrochemical biosensors in North America. Biosens. Bioelectron. **172**, 112750 (2021)
57. R.C. Chiechi et al., Eutectic gallium–indium (EGaIn): A moldable liquid metal for electrical characterization of self-assembled monolayers. Angew. Chem. Int. Ed. **47**(1), 142–144 (2008)
58. M.D. Dickey, Stretchable and soft electronics using liquid metals. Adv. Mater. **29**(27), 1606425 (2017)
59. D. Wang et al., Chemical formation of soft metal electrodes for flexible and wearable electronics. Chem. Soc. Rev. **47**(12), 4611–4641 (2018)
60. N. Kurra, G.U. Kulkarni, Pencil-on-paper: electronic devices. Lab Chip **13**(15), 2866–2873 (2013)
61. M. Zabihipour et al., High yield manufacturing of fully screen-printed organic electrochemical transistors. npj Flex. Electron. **4**(1), 1–8 (2020)
62. C. Koutsiaki et al., Efficient combination of Roll-to-Roll compatible techniques towards the large area deposition of a polymer dielectric film and the solution-processing of an organic semiconductor for the field-effect transistors fabrication on plastic substrate. Org. Electron. **73**, 231–239 (2019)
63. S. Chung, K. Cho, T. Lee, Recent progress in inkjet-printed thin-film transistors. Adv. Sci. **6**(6), 1801445 (2019)
64. B. Ong, Semiconductor ink advances flexible displays. Laser Focus World **40**(6), 85–86 (2004)
65. G. Grau et al., Gravure-printed electronics: Recent progress in tooling development, understanding of printing physics, and realization of printed devices. Flex. Print. Electron. **1**(2), 023002 (2016)

66. C. Linghu et al., Transfer printing techniques for flexible and stretchable inorganic electronics. npj Flex. Electron. **2**(1), 1–14 (2018)
67. H.-J. Kim-Lee et al., Interface mechanics of adhesiveless microtransfer printing processes. J. Appl. Phys. **115**(14), 143513 (2014)
68. H. Kozuka et al., Ceramic thin films on plastics: A versatile transfer process for large area as well as patterned coating. ACS Appl. Mater. Interfaces **4**(12), 6415–6420 (2012)
69. T.A. Pham et al., A versatile sacrificial layer for transfer printing of wide bandgap materials for implantable and stretchable bioelectronics. Adv. Funct. Mater. **30**(43), 2004655 (2020)
70. A. Shivayogimath et al., Do-it-yourself transfer of large-area graphene using an office laminator and water. Chem. Mater. **31**(7), 2328–2336 (2019)
71. M.S. Mannoor et al., Graphene-based wireless bacteria detection on tooth enamel. Nat. Commun. **3**(1), 763 (2012)
72. H.-U. Kim et al., Flexible MoS2–polyimide electrode for electrochemical biosensors and their applications for the highly sensitive quantification of endocrine hormones: PTH, T3, and T4. Anal. Chem. **92**(9), 6327–6333 (2020)
73. M. Marchena et al., Dry transfer of graphene to dielectrics and flexible substrates using polyimide as a transparent and stable intermediate layer. 2D Mater. **5**(3), 035022 (2018)
74. K. Sharma, A. Arora, S.K. Tripathi, Review of supercapacitors: Materials and devices. J. Energy Storage **21**, 801–825 (2019)
75. Y.-Z. Zhang et al., Printed supercapacitors: Materials, printing and applications. Chem. Soc. Rev. **48**(12), 3229–3264 (2019)
76. C. Gao et al., A seamlessly integrated device of micro-supercapacitor and wireless charging with ultrahigh energy density and capacitance. Nat. Commun. **12**(1), 1–10 (2021)
77. J. Pu et al., Highly flexible MoS2 thin-film transistors with ion gel dielectrics. Nano Lett. **12**(8), 4013–4017 (2012)
78. D. Bhatt, S. Kumar, S. Panda, Amorphous IGZO field effect transistor based flexible chemical and biosensors for label free detection. Flex. Print. Electron. **5**(1), 014010 (2020)
79. J. Ping et al., Scalable production of high-sensitivity, label-free DNA biosensors based on back-gated graphene field effect transistors. ACS Nano **10**(9), 8700–8704 (2016)
80. S. Xu et al., Real-time reliable determination of binding kinetics of DNA hybridization using a multi-channel graphene biosensor. Nat. Commun. **8**(1), 1–10 (2017)
81. N. Kumar et al., Dielectrophoresis assisted rapid, selective and single cell detection of antibiotic resistant bacteria with G-FETs. Biosens. Bioelectron. **156**, 112123 (2020)
82. N. Kumar et al., Detection of a multi-disease biomarker in saliva with graphene field effect transistors. Med. Dev. Sens. **3**, e10121 (2020)
83. A.A. Cagang et al., Graphene-based field effect transistor in two-dimensional paper networks. Anal. Chim. Acta **917**, 101–106 (2016)
84. C. Huang et al., An integrated flexible and reusable graphene field effect transistor nanosensor for monitoring glucose. J. Mater. **6**(2), 308–314 (2020)
85. Z. Wang et al., A flexible and regenerative aptameric graphene–Nafion biosensor for cytokine storm biomarker monitoring in undiluted biofluids toward wearable applications. Adv. Funct. Mater. **31**(4), 2005958 (2021)
86. M. Kim et al., All-parylene flexible wafer-scale graphene thin film transistor. Appl. Surf. Sci. **551**, 149410 (2021)
87. A.M. Hussain et al., Ultrastretchable and flexible copper interconnect-based smart patch for adaptive thermotherapy. Adv. Healthc. Mater. **4**(5), 665–673 (2015)
88. P. Mostafalu et al., Smart bandage for monitoring and treatment of chronic wounds. Small **14**(33), 1703509 (2018)
89. D.-H. Kim et al., Epidermal electronics. Science **333**(6044), 838–843 (2011)
90. A.V. Mohan et al., Merging of thin-and thick-film fabrication technologies: Toward soft stretchable "island–bridge" devices. Adv. Mater. Technol. **2**(4), 1600284 (2017)
91. C. Votzke et al., 3D-printed liquid metal interconnects for stretchable electronics. IEEE Sensors J. **19**(10), 3832–3840 (2019)

92. Z. Huang et al., Three-dimensional integrated stretchable electronics. Nat. Electron. **1**(8), 473–480 (2018)

93. J.-K. Song et al., Nanomaterials-based flexible and stretchable bioelectronics. MRS Bull. **44**(8), 643–656 (2019)

94. H. Lee et al., An endoscope with integrated transparent bioelectronics and theranostic nanoparticles for colon cancer treatment. Nat. Commun. **6**(1), 10059 (2015)

Chapter 2
Sensors and Platforms for Flexible Bioelectronics

2.1 Introduction

There have been tremendous progresses made in the area of human machine interface, electroceutical therapy, and wearable diagnostics [1, 2]. This chapter will cover flexible bioelectronics platforms for biomedical applications including sensing and therapeutic delivery. An overview of flexible sensors will be provided including biophysical and biochemical sensors. This is followed by two case studies where a smart bandage platform for management of wound healing and a thread-based diagnostic/therapeutic platform will be introduced.

2.1.1 Flexible Biosensors

In the past decades, there was a growing interest on using wearable biosensors to monitor one's health conditions. The global wearable sensor market is expected to grow 30% from 2019 to 2024 [3]. Conventional electronic sensors are made from semiconducting materials including silicon and metal oxide, which are rigid and inflexible. The need for flexibility emerged due to the requirement of intimate and stable interfaces with human body (such as conformal contact with the skin and matchable mechanical properties with the tissue). Recent research development has been focused on using flexible, stretchable, lightweight, and biocompatible materials for making wearable sensors, as well as efforts to miniaturize electronics and circuits for readout and data transmission [4, 5].

Dr. Junfei Xia (Tufts University) and Dr. Sameer Sonkusale (Tufts University) contributed to this chapter.

© The Author(s), under exclusive license to Springer Nature Switzerland AG 2022
S. Sonkusale et al., *Flexible Bioelectronics with Power Autonomous Sensing and Data Analytics*, https://doi.org/10.1007/978-3-030-98538-7_2

Wearable biosensors are designed to measure vital signs including heart rate, body temperature, blood pressure, respiration rate, as well as biomarkers from body fluids such as glucose, lactate, and electrolytes. These biomarkers function as direct or indirect indicators of one's health condition and need to be monitored in long term in order to establish personalized references for personalized therapy. Biomarkers are classified into biophysical markers and biochemical markers. Biophysical markers include vital body signs such as electrocardiogram (ECG), pulse rate, and other electrophysiological signals. Biochemical markers are soluble biomolecules found in body fluids such as saliva, blood, sweat, tear, and urine, or volatile organic compound found in breath. A detailed overview of biophysical and biochemical markers will be provided later in this chapter.

2.1.2   Sensing Principle

Sensor is a device that can detect changes or events in its environment. The basic units of a biosensor are receptor and transducer. The receptor comes in close contact with the analyte and the chemical/physical interactions between the two are translated into a measurable signal by the transducer (Fig. 2.1). The receptor, or the recognition unit, usually consists of a layer of biomolecules such as antibody, enzyme, nucleic acid, or cells that can selectively interact with the target biomarker. The transducer can be optical, electrochemical, thermal, or piezoelectric, depending on how the signal is measured. In the view of the whole sensing device, the transducer signal is further amplified by the electronic circuitry and processed using computer algorithms to present the end results to the user. The information on circuits design and signal processing related to sensors will be discussed in the later chapters.

2.1.2.1   Optical Transducer

An optical transducer converts receptor–analyte interaction into photons and outputs and optical signal. It is widely used in developing wearable biosensors due to their

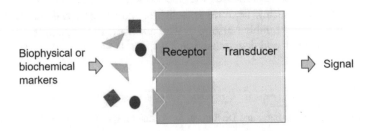

Fig. 2.1 Basic components of sensor

high sensitivity, noninvasiveness, high spatiotemporal resolution, and low cost. As this chapter mainly focuses on bioelectronic platforms, readers are referred to a more comprehensive review on wearable optical biosensor [6]. It is to be noted that optical approaches that typically involve bulky and rigid optical components are less preferred over electrochemical approaches discussed next.

2.1.2.2 Electrochemical Transducer

An electrochemical transducer converts receptor-analyte interaction into electrons and outputs an electronic signal. In a typical electrochemical sensor, a minimum of two electrodes are employed to either oxidize or reduce the analyte of interest. These electrodes can be easily realized on flexible substrates using the approaches discussed in the previous chapter. The generated current that is produced from this reaction is measured as an indication of the analyte concentration. Based on the method of analysis, electrochemical sensing techniques can be divided into two major types: static method and dynamic method. In the static method, a measurable potential or charge accumulation is generated by the electrode reaction, such as potentiometry widely used in ion-selective electrodes [7, 8]. It is featured by high selectivity and fast response time. In dynamic methods, a current flows through the electrodes at the application of either a constant or varying potential between the electrodes. It contains a wide variety of commonly used techniques including cyclic voltammetry (CV) and chronoamperometry (CA). A detailed classification of various electrochemical analysis techniques is shown in Fig. 2.2. Dynamic methods are featured by high sensitivity, broad dynamic range, and small sample volume. Some other commonly used techniques include differential pulse voltammetry (DPV), anodic stripping voltammetry, and electrochemical impedance spectroscopy (EIS). Here we would like to highlight several techniques as they have been applied in the development of some very successful biosensors. Amperometry is almost

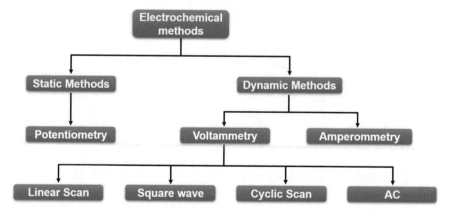

Fig. 2.2 Different modes of electrochemical analytic techniques

ubiquitously used in enzyme-based biosensor such as the glucose sensor as the generated current due to the oxidation of glucose at a fixed redox potential is proportional to the concentration of glucose [9]. In EIS, one measures the impedance at the electrode and electrolyte interface. A binding activity of a target analyte to a target-specific receptor immobilized on the electrode alters the charge transfer across the electrode–electrolyte interface, and it is measured using impedance across a range of frequencies. Both DPV and EIS are broadly used in affinity-based immunosensors as they provide real-time and sensitive detection of analytes in ultra-low quantity.

2.1.3　Thermal Transducer

A thermal transducer converts receptor–analyte interaction into thermal energy and output change in temperature. In most cases, a thermistor is used to convert change in temperature into an electronic signal for processing and readout. Thermal transducer is widely used in enzyme-based biosensors as almost all enzyme-catalyzed reactions release heat, where the ratio of substrate used to generate heat is proportional to the amount of heat [10]. Thermal biosensors are widely applied to monitor bioprocesses such as fermentation process. The analyte of interests includes temperature, pressure, pH, partial pressure of oxygen (pO_2), concentration of metabolic products and derived by-products, and so on [10].

2.1.4　Piezoelectric Transducer

Piezoelectric transducer, by definition, converts mechanical stress into electronic signal. There are two types of piezoelectric transducers used in biosensing. For measuring biochemical molecules, or biomarkers, an alternating voltage (AC) is applied on the surface of electrodes which causes mechanical oscillation of the piezoelectric material. The binding of analytes to the electrodes results in change of the oscillation frequency that can be measured and is proportional to the mass bound. This type of piezoelectric transducers is suitable for the construction of affinity-based sensors without the need to apply any specific reagents [11, 12]. Another type of piezoelectric transducer is used for measuring physiological signals such as heart rate, respiration rate, and body-movement signal. In these sensors, the deformation of the piezoelectric materials caused by pulse or respiration is directly converted to electronic signals and measured. This type of piezoelectric sensors is widely used in wearable sensors [13, 14]. Most piezoelectric transducers use bulk and rigid elements and are therefore not quite amenable for flexible bioelectronics application.

2.2 Characteristics of Biosensors

It is important that we also discuss certain key features of all types of biosensors. These are also the key metrics for evaluating the performance of a biosensor.

Limit of Detection

Limit of detection (LOD) refers to the minimum amount or concentration of the analyte that can be detected. It is one of the most important metrics of a biosensor as many disease-related diagnosis requires very low LOD in the range of ng/mL or even pg/mL. A low LOD is critical for detecting trace amount of analytes present in body fluids which is especially important for early diagnosis.

Detection Range

Detection range is a reflection of how much change caused by the external stimuli can be measured. In biochemical sensors, it is the concentration range between upper and lower limit of detections. A typical response curve of a biosensor is shown in Fig. 2.3 This sigmoid curve is characterized by the lower limit of detection (LLOD), upper limit of detection (ULOD), slope of the curve (sensitivity), and linear range. The signal gets diminished when the concentration is close to LLOD and saturated when the concentration is close to ULOD. In most cases, a broad linear range is preferred to cover a wide working condition.

Sensitivity

The sensor's sensitivity is defined as the ratio between the output signal and the input measured property. For example, a pressure sensor can have a sensitivity of 500 mV/psi. In general, a high sensitivity is preferred since it determines the smallest change that can be detected by the sensor.

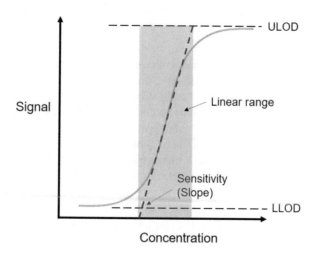

Fig. 2.3 Typical response curve of a biosensor. Blue line is sensor's response

Selectivity

Selectivity is a term used more frequently in biochemical sensors. It indicates the sensor's ability to differentiate one analyte from another. It is the most important factor to consider when choosing the type of bioreceptor. Antibody–antigen interaction and enzymatic recognition are traditionally used in bioreceptor design as they exhibit high selectivity toward certain type of analytes. In recent years, aptamer has emerged as a new class of bioreceptor with high binding affinity due to the SELEX process where binding with the target analyte has already been validated.

Reproducibility

Reproducibility refers to the ability of the sensor to generate identical results when it is measuring the same level of external stimuli. This requires the sensing procedure to be reversible (i.e., physical or chemical processes that happen during the sensing step do not change the sensor itself). Reproducibility can be affected by both bioreceptors and transducers. For instance, the stability of immobilized enzyme on surface of the sensing electrodes could influence the rate of enzymatic reaction and thus the detected current level. The accuracy of electronics in the transducer could affect the signal-to-noise ratio and thus the reproducibility of measurements.

Stability

Stability of a biosensor can be defined as the ability to maintain the same level of output signal under various environmental disturbances. Again, this can be contributed from two parts: the bioreceptor and the transducer. In biomedical application, stability is a very crucial criteria to consider especially when the sensor needs to be functioning for a significant period of time. When interfacing with the biological system, there are issues such as bio-fouling and enzyme/antibody degradation that will decrease the stability of a biosensor. In most cases, the transducer electronics are considered much more stable compared with the actual sensing part.

Biocompatibility

Biocompatibility is defined as the level of compatibility of the material/device with living tissue. From a broad point of view, this includes toxicity of materials, reactivity with living tissue, mechanical compatibility, and biodegradability. Traditional silicon and metal oxide-based semiconducting devise are not biocompatible since they do not match the Young's modulus of soft tissues. The flexible substrates such as PDMS, polyimide, and PET have similar mechanical properties with tissue and are bio-inert, which are more biocompatible. For wearable biosensors, the requirement of biocompatibility is lower than implantable sensors as they stay only on top of skin, where flexibility is the priority factor. One example could be the use of nanomaterials (such as carbon nanotubes and metal nanoparticles) in fabricating flexible biosensors which could cause cytotoxicity when implanted.

2.2.1 Physical Sensors

Physical sensors are used to measure biophysical markers. They usually need to get in close contact with the body part where the signal is generated. This section will overview the list of biophysical markers including vital signs and biopotential signals with examples of flexible sensors for measuring them.

2.2.1.1 Physical Sensors for Monitoring Vital Signs

Vital signs are core medical signs that indicate the state of body's functions. The primary vital signs include body temperature, heart rate and heart rate variability, respiration rate, and blood pressure.

Body Temperature Body temperature is a key parameter associated with health. It varies by person, age, and activity but is generally maintained in a normal range. An elevated body temperature could be an indication of fever caused by infection or illness. A decreased body temperature could be caused by the environment, use of drugs, or diseases such as stroke and sepsis [15]. Flexible temperature sensors can be made from temperature-sensitive conductive materials whose electrical resistivity changes with the temperature. One example is the doping of flexible substrates with conductive fillers such as carbon black, carbon nanotube, and graphite to make flexible, conductive composite materials [16–18]. Another example is the transfer printing of conducting materials and patterns onto a flexible substrate such as PDMS (Fig. 2.4a) [19, 20].

Heart Rate and Heart Rate Variability Heart rate (HR) is an important biomarker for assessing cardiorespiratory fitness and diagnosing cardiovascular diseases [21, 22]. Heart rate variability (HRV), the variation of time interval between heartbeats, is also frequently used to track well-being and assess overall cardiac health [23]. HR and HRV are traditionally measured by electrocardiograms (ECG) and photoplethysmography (PPG). Most wearable HR sensors are based on piezoelectric transducers that convert deformation of the skin caused by pulse wave to an electric signal. In recent years, researchers have developed flexible sensors aiming to provide high accuracy, high sensitivity as well as comfort of wearing for measuring HR and HRV. Sekine et al. reported a HR sensor made from piezoelectric polymer that is compact, flexible, and with high pressure sensitivity, where the deformation of skin caused by pulse is measured by the sensor [24]. Yoon et al. developed a skin-mounted sensor patch for measuring skin temperature, skin conductance, and HR [25]. The sensor possesses the size of stamp in order to enhance the wearing comfort. The use of flexible piezoelectric material, P(VDF-TrEE), and a supporting polyimide layer provides additional flexibility of the patch. The sensing patch is also suitable for monitoring HRV as demonstrated by the authors.

Respiration Rate Breath monitoring plays an essential role in the assessment of human health conditions. Respiration rate (RR) can serve as an indicator of mental diseases and mental disorders [26]. Changes in the RR and depth of respiration

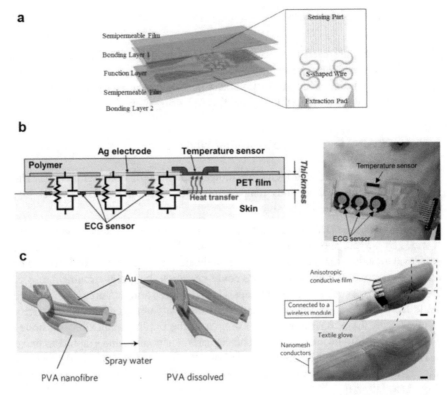

Fig. 2.4 Biophysical sensors. (**a**) Illustration of the design of a biocompatible and stretchable temperature sensor. (**b**) Left: Equivalent circuit of the ECG sensor and heat transfer through a PET film from skin to the device. Right: Photo of the fabricated device. (Adapted with permission from Ref. [36]). (**c**) On-skin nanomesh electronics and the depiction of an on-skin wireless sensor system. (Adapted with permission from Ref. [37])

are associated with asthma, obstructive sleep apnea, heart failure, heart attack, and lung cancer [27, 28]. The techniques for measuring RR are based on measuring chest/body movement, respiratory airflow, respiratory sound, air temperature and humidity, and others. One of the most common and direct methods is to measure humidity from inhaled and exhaled air. For real-time monitoring, the humidity sensors can be integrated into a face mask. For instance, Li et al. reported a flexible humidity sensor for measuring RR based on silk fabrics [29]. The sensing layer was fabricated by spray-coating of graphene oxide (GO) onto the silk fabrics. The water molecules in the environment interact with the surface functional groups of GO and result in a decrease of electrical impedance, which is measured for differentiating the humidity change during inhalation–exhalation cycles. Flexible strain sensors that measure chest/body movement caused by breathing also attract significant attention as they are highly sensitive and accurate and rarely affected by surrounding environment such as fluctuation in air humidity or temperature. Park et al. developed a flexible capacitive pressure sensor that can be placed on the waist belt to monitor respiration [30]. The sensor was made from a porous Ecoflex sheet sandwiched between two flexible electrodes. The electrodes were fabricated with doping PDMS with conductive silver nanowires (AgNWs) and carbon fibers. The compression of the porous Ecoflex (dielectric layer) due to stomach inflation caused change in the distance between two electrodes thus changing the capacitance value.

Blood Pressure Blood pressure (BP) is one of the most important vital signs in the cardiovascular system. It is responsible for delivering oxygen and nutrients to the whole body through blood circulation. Abnormal levels of BP can be indications of various diseases. For example, high BP is indicative of likelihood of strokes, heart attack, and heart failure. Low BP is indicative of cardiogenic shock, sepsis, and hemorrhage. The early diagnosis of these diseases relies on the long-term, continuous monitoring of BP. Conventionally, BP is measured using the auscultatory method which uses a mercury sphygmomanometer to determine the systolic BP (SBP) and diastolic BP (DBP). However, this method requires a well-trained physician and is not suitable for continuous monitoring. As commercial automatic BP meters become available, self-recording of BP at home is possible but still not applicable to long-term monitoring due to the bulky size of the inflatable cuff and the high power consumption of such devices. The recent research development on flexible skin-mounted BP sensors with low power consumptions (nW–μW) have enabled continuous recording of BP together with other bioelectric signals such as heart rate and ECG. A cuffless BP sensor integrated with an epidermal ECG sensor was developed by Luo et al. [31]. The sensor was fabricated by lithographically patterned gold electrode on the polyimide substrate and adhered to skin using PDMS tape. The stretchability and flexibility were provided by the serpentine designed as proposed by the Rogers' group [32]. A carbon-decorated fabric was laid on top of gold electrodes to offer piezoresistive behavior of the device that converts pulse and pressure signal into a current or voltage output. The SBP and DBP were calculated based on the pulse transient time (PTT) measured from ECG. In addition, the sensor requires much lower power consumption (3 nW) which is a general advantage of piezoelectric sensors in which the sensor elements can be self-powered.

2.2.1.2 Physical Sensors for Measuring Biopotential Signals

Biopotential signals are generated by the collective activities of electrogenic cells, such as neuron, muscle, and cardiac cells. These signals are measured by wet electrodes which consist of an adhesive layer, a gel layer, and the underlying metal electrode. However, wet electrodes suffer from various issues including drying of the gel, discomfort while wearing and peeling off the electrodes, and possible allergic reactions caused by the gel layer. Flexible substrates such as soft polymers and textiles have shown great potential as recording electrodes for such applications. PDMS is most commonly used as the substrate for dry electrodes as it is flexible, transparent, permeable to air and moisture, and biocompatible to human skin.

To fabricate the electrodes, metals (Ti and Au) have been patterned on PDMS to create flexible electrodes that can be worn on the forearm for measuring ECG [33]. Wang et al. developed a flexible dry electrode made of PDMS coated by sputtered gold. The electrodes with pin structures ensure better contact on hair sites for measuring electroencephalography (EEG) [34]. Alternatively, PDMS can be doped with other conductive materials such carbon nanotubes to create PDMS/CNT composite which excels in both flexibility and electrical conductivity. This composite material was used as electrodes for long-term monitoring of ECG [35]. As skin–electrode contact is important for good conductivity, Yamamoto et al. developed a high adhesive conductive polymer by mixing ethoxylated polyethylenimine (PEIE) and CNT with PDMS and used it as the adhesion layer between skin and the electrode (Fig. 2.4b) [36]. Recently, electrospun nanofibers were applied to fabricate nanomesh electrodes for on-skin sensing applications. The ultrathin, lightweight, and air-permeable nanomesh was mounted on skin for measuring electromyography (EMG) signals (Fig. 2.4c) [37]. Textile-based electrodes are also attractive as they are intrinsically flexible and can be seemingly integrated with clothing. Screen printing and other coating techniques are used to make textile conductive. They can be embedded using textile weaving and embroidering techniques. To date, textile-based electrodes have been applied in the acquisition of various biopotential signals including ECG, EEG, EMG, and electrooculography (EOG). Readers are referred to a more detailed review article on textile-based electrodes for biopotential monitoring [4].

2.2.1.3 Pressure and Strain Sensor

Pressure and strain sensors are used to measure body movements such as running and walking. They can also monitor subtle skin movements such as phonation of words and saliva swallowing. In addition, highly sensitive strain sensors have been used to measure vascular dynamics such as pulse wave pressure, pulse wave velocity, and blood pressure (covered in the previous section) [38]. Two types of pressure and strain sensors are commonly used: resistive- or capacitive-based sensors. In resistive-based sensors, stretching changes resistance of the conductive substrates resulted from changes in their length and geometry. In capacitive-based sensors, two conductive substrates are separated by a dielectric layer to form a capacitor. When stretched or compressed, the deformation changes the dimension of the device which leads to the change of capacitance. A wide variety of flexible substrates have been reported to fabricate pressure and strain sensors over the past decades. These include but not limited to PDMS, polyurethane, PET, polyimide, elastomeric polymers, and textile. In order to function as sensors, these substrates need to be rendered conductive, usually through integration of conductive materials such as graphene, carbon nanotubes, AgNW, and conducting polymers. Approaches discussed in Chap. 1 such as dip coating, screen printing, and electro-/electroless plating have enabled low-cost and cleanroom-free fabrication of these sensors (Fig. 2.4).

2.2.2   Chemical/Biological Sensors

Biochemical sensors rely on chemical reactions or receptor–ligand binding to detect the presence of target analytes. By their nature, biochemical markers are biomolecules or cells including protein, cytokines, DNA, RNA, ions, metabolites, dissolved gases, and circulating tumor cells. They are found in body fluids which are either enclosed inside body such as blood and interstitial fluid or secreted by body such as sweat and urine. Compared with biophysical sensors, biochemical sensors are intrinsically challenged by the capability to do real-time sensing due to sampling issue. Below we will review clinical-relevant body fluids for disease diagnosis and provide several examples of flexible biosensor platform with a focus on their sampling and sensing mechanism. Table 2.1 summarizes the characteristics of major types of body fluids for sensing.

Blood   Blood accounts for about 7% of the total body weight. It contains the most abundant biomarkers across the whole body. Blood tests are routinely conducted in clinics and hospitals and blood-based diagnostics are considered as "gold standards" among all type of tests. However, blood sampling is invasive with the risk of infection if not performed properly. It is usually not applicable for real-time monitoring as most blood tests are single-point measurement with long turnaround time. In addition, blood is a heterogenous fluid containing blood cells, plasma, platelets, and biomolecules. It needs to be processed (e.g., separation, sedimentation, purification) before biochemical analysis. Due to these limitations, blood-based biosensors are mostly for off-body measurement. Recent research advancement in fingerstick-based blood testing, including phenylalanine [39] and cardiac troponin [40], only requires a fingerstick which may hold promise for point-of-care blood analysis [41].

Table 2.1   Summary of major types of body fluids for sensing application

Body fluids	Disease-related biomarkers	Invasiveness of sampling
Blood	Contains the most comprehensive panel of biomarkers Some biomarkers are serum-bound	Invasive
Sweat	Strong correlation with blood concentration of small molecules	Non-invasive
Interstitial fluid	Contains most of the blood biomarkers but in less concentration	Minimally Invasive
Urine	Concentration of biomarkers vary from person to person High ionic strength may confound detection	Noninvasive
Saliva	Significantly lower concentration of biomarkers Some biomarkers are in free form not serum-bound (e.g., cortisol)	Noninvasive
Breath	Volatile organic compounds Correlation between breath markers and few select diseases has been strong	Noninvasive

Sweat Sweat is secreted by sweat gland which contains a large array of biomarkers. More than 800 unique proteins have been identified in sweat of which 57 are proteases and 38 are protease inhibitors [42]. The concentrations of small molecules in sweat are strongly correlated with those in the blood, while there is not enough evidence to show the same for large molecules [43]. Sweat monitoring is noninvasive with the huge potential for real-time and continuous measurement. Several challenges associated with sweat sensing include the following (1). Very low concentrations of biomarkers compared with those in blood. (2). Small amount of sweat sample that can be collected and evaporation. (This has been partially addressed recently by microfluidic sampling technology) [44]. (3). Potential contamination by other entities present on the skin. Despite these challenges, researchers have developed wearable sweat sensors that can reliably detect glucose, lactate, electrolytes, cytokines, etc., in real time. Gao et al. developed a flexible and fully integrated sweat sensor array for multiplexed detection of metabolites, electrolytes, and skin temperature [45]. The sensor array consists of immobilized glucose oxidase (GOx) and lactate oxidase (LOx) and ion-selective membranes on printed electrodes connected with conductive traces. The enzymes generate a current readout due to the catalytic reaction when glucose or lactate are present in sweat, while the current is in proportional to their concentrations. The ion-selective electrodes attract specific ions which generate a voltage readout. A flexible printed circuit board (FPCB) was connected to the array and communicates wireless to a smartphone for readout. The whole device is wearable as a "smart wristband" as shown in Fig. 2.5a.

Interstitial Fluid Interstitial fluid (ISF) is a type of body fluids existing between blood vessels and cells. It transports nutrients and wastes between cells and capillaries. It contains a lot of biomolecules including sugar, salts, fatty acids, enzymes, neurotransmitters, cell metabolic products, and many others. It is currently used in clinic for continuous glucose monitoring. Compared with blood, it is relatively difficult to sample from the body using conventional syringe needles. In recent years, the development of microneedles has enabled more facile sampling of ISF as well as painless delivery of medicine including vaccine and drugs [46–48]. In addition, microneedles were used as sensing electrodes for electrochemical analysis of ISF [49]. In another proof-of-concept study, reverse iontophoresis was integrated with flexible electrochemical sensors and assembled as a "skin-tattoo" for ISF glucose monitoring [50]. The "tattoo" consisted of a pair of iontophoresis electrodes responsible for extraction of ISF glucose, and a working and counter/reference electrode modified by Prussian blue and GOx for glucose sensing (Fig. 2.5b). This technology showed considerable promise for noninvasive monitoring of biomarkers in ISF.

Urine Urine contains 95% of water and is a commonly used body fluid for lab testing including drug test and kidney function. It contains protein, DNA, and metabolites that are specific indicators to some diseases such as bladder cancer. Urine is easy to collect and process and is usually in large volume. However, the high ionic strength in urine can confound detections using many electrochemical and

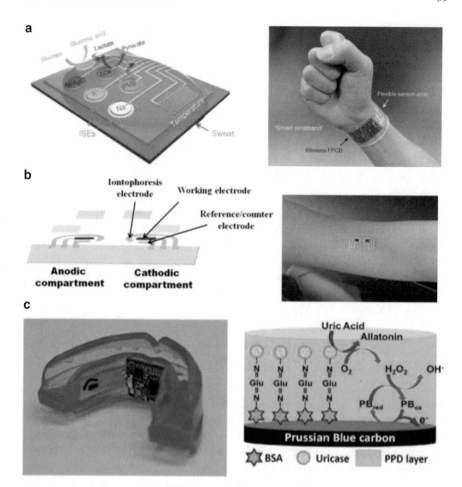

Fig. 2.5 Biochemical sensors. (**a**) Flexible wearable electrode array for sweat analysis. Adapted with permission from ref [45]. (**b**) Tattoo-based platform for noninvasive glucose sensing. (**c**) Left: Photograph of the mouthguard biosensor integrated with wireless amperometric circuit board. Right: Reagent layer of the chemically modified printed Prussian-blue carbon working electrode containing uricase for SUA biosensor. (Adapted with permission from Ref. [53])

electronic approaches. In addition, the concentration of biomarkers in urine varies from person to person thus making it more challenging for accurate measurement. Urine sensing is noncompliance with long term and real time due to the social stigma associated with it.

Saliva Saliva is secreted by salivary glands in the mouth which consists of mostly water (98%) and electrolytes, mucus, enzymes, growth factors, and antibacterial compounds. Saliva contains clinically relevant biomarkers such as C-reactive protein (CRP) and cortisol that are useful for disease diagnosis, namely, cardio-vascular disease and renal disease [51]. Saliva is easy for collection, storage, and transportation. In recent years, it has drawn particular interest for point-of-care diagnostics due to the abovementioned advantages [52]. Kim et al. developed a saliva uric acid sensor based on mounting the sensor and readout circuits on a mouthguard (Fig. 2.5c) [53]. The sensing electrodes were screen printed on flexible polyethylene terephthalate (PET) sheet and modified by Prussian blue as electron mediator and uricase as recognition component for uric acid. An anti-fouling layer consisting of polymerized o-phenylenediamine (PPD) was deposited to stabilize the electrodes in biological fluids. The sensor was able to wirelessly transmit data in real time to a smartphone or laptop owing to the full integration of miniaturized electronics including a potentiostat, microcontroller, and a Bluetooth Low-Energy transceiver.

Breath Breath is a special type of "body fluid" which is a mixture of thousands of volatile organic compounds (VOCs) and microscopic droplets originating from lungs and airways. Biomarkers found in breath are relevant to pulmonary diseases, cardiovascular diseases, cancer, cystic fibrosis, rheumatoid arthritis, etc. Breath-based diagnostics are noninvasive and straightforward. It is also readily available for real-time sensing with the development of novel breath sampling devices and gas-based biosensors [54, 55].

2.2.3 Fully Integrated Wearable Bioflexible Platform

Recent advances in flexible electronics, soft microfluidics, and wireless power transfer have led to the development of fully integrated, self-contained wearable biosensors that are flexible, lightweight, and compatible for on-skin attachment. In these platforms, the system is fully integrated into thin patches or wearable wristband including the sensing electrodes (directly interfacing with skin), signal reading and processing unit, data transmission module, and power supply. The integrated circuits for signal processing are either embedded in a flexible substrate and interconnected by serpentine patterned conductive traces or consolidated on a flexible printed circuit board (Fig. 2.6a, b). Alternatively, intrinsically flexible materials are used to fabricate a transistor array with high array density that is promising to make integrated circuit elements (Fig. 2.6c).

Chung et al. developed an epidermal patch for health monitoring of newborns in neonatal intensive care unit [56]. As shown in Fig. 2.6a, chip-scale integrated circuit components are embedded in PDMS as a flexible substrate. The stretchability was provided by connecting these chips with serpentine copper traces that allows up to 16% stretching. Filament metal mesh microstructures with fractal geometry were used as sensing electrodes to record ECG signal. A red LED and the sensing circuitry were embedded in the second patch for measuring photoplethysmograms (PPGs). In addition, both patches contain a magnetic loop antenna for both wireless data transmission from the sensor and wireless power delivery to drive the sensing circuitry. The two epidermal patches were validated in both healthy and premature

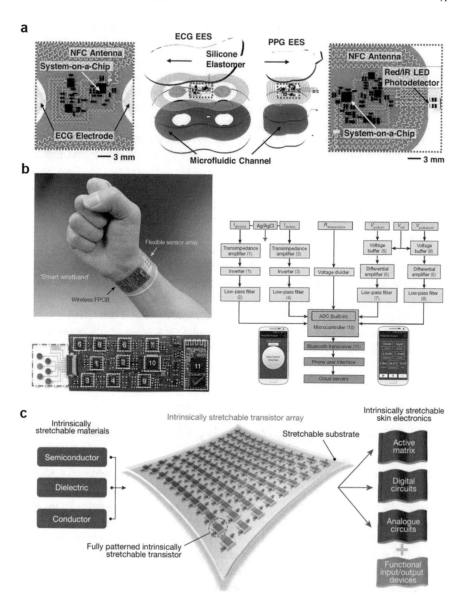

Fig. 2.6 Fully integrated bioflex platforms. (**a**) Schematic illustration of ultrathin, skin-like, wireless, battery-free modules for recording ECG and PPG data and skin temperature for use in the neonatal intensive care unit. (**b**) Image and illustration of the flexible integrated sensing array (FISA) for multiplexed perspiration analysis. (Adapted with permission from Ref. [45]). (**c**) Intrinsically stretchable transistor array as a core platform for functional skin electronics. (Adapted with permission from Ref. [60])

infants, and the results were comparable to clinical standards. A second example is illustrated by Gao et al. where they developed a fully integrated sensor array for sweat monitoring [45]. A prototype wristband sensor is made which is completely free from external analysis (Fig. 2.6b). The electrode array was fabricated on a flexible PET film and was capable of multiplexed sensing of metabolites and electrolytes in sweat. In addition, the use of Prussian blue as a redox mediator minimized reduction potential to 0 V thus enabling power-free driving of the sensing circuitry. This work used flexible printed circuit board (FPCB) technology that can mount commercially available electronics for signal processing and data transmission onto a flexible plastic board.

Intrinsically flexible materials have been proposed to fabricate electronic circuit elements such as transistors [57–59], but the lack of scalability in fabrication and reliability of their functions have hindered their application. Recently, a scalable fabrication approach for intrinsically flexible transistor array was proposed by Wang et al. [60]. In their process, a sacrificial layer was firstly deposited on a silicon wafer. Then stretchable dielectric, stretchable semiconductor, and conductor materials were patterned to create the transistor structure. The sacrificial layer was finally dissolved to release the whole transistor array. This process is scalable to the making of 347 transistors per square centimeter. Functional test showed that a skin-mounted sensor consisted of a stretchable pulse sensor and a stretchable amplifier (built from flexible transistors) was able to detect physiological signals such as pulse. This demonstrates that intrinsically flexible transistors could be the core building blocks of skin electronics (Fig. 2.6c).

2.2.4 Smart Bandages for Wound Healing and Therapy

Wound healing is a complicated physiological process which undergoes inflammatory, proliferative, and tissue remodeling phases where various types of cells, extracellular matrix, and signaling molecules are present at the wounded sites [61]. Normally the wound heals at a predictable and manageable rate such as acute wound. However, an uncoordinated and self-sustained phase of inflammation could develop that leads to chronic wound if the stepwise process is disturbed [62, 63]. It is estimated that 5.7 million Americans are suffering from chronic wounds that cost at least $28 billion to the healthcare system [64]. In most cases, infection is not detected onset but after additional tissue damage has occurred. It is estimated that 25% of infected chronic wounds lead to amputation. Chronic wounds are the leading cause of nontraumatic limb amputation. In this section, we will highlight the work on smart bandage for wound healing and therapy developed by the Sonkusale lab at Tufts University. Rigid solutions relying on hard components to monitor wounds will clearly hamper the healing process, and there is a clear need for flexible bioelectronic wound dressing to monitor and accelerate healing. As shown in Fig. 2.7, the team developed a smart bandage can monitor biomarkers for wound healing and provide real-time drug delivery. The system is featured

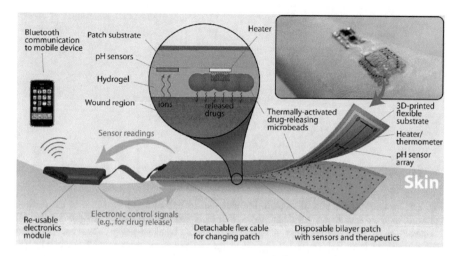

Fig. 2.7 Smart bandage for wound monitoring and therapy. (Adapted with permission from Ref. [71])

by tissue engineered dressing, integrated miniaturized sensors, on-demand drug delivery, electrochemical potentiostat, wireless transmission, and power telemetry, all of which are in a compact design and can be readily attached on the skin.

2.2.4.1   Smart Bandage for Wound Monitoring

The ability for continuous wound monitoring is essential for wound management. Oxygen is ubiquitously involved in all steps of wound healing due to increased energy demand resulting from cell proliferation, inflammation response, and collagen synthesis [65, 66]. The lack of oxygen supply results in local tissue hypoxia and leads to hampered tissue repair. The evaluation of tissue oxygenation can therefore be used as an indication in the management of acute and chronic wounds. More importantly, a wearable device for monitoring tissue oxygenation in wound bed would be of great value in the management of acute and chronic wound. Mostafalu et al. developed a wearable oxygen sensing patch made on flexible parylene substrate with wireless connectivity for smartphone readout [67]. The sensing of oxygen is realized by deposition of zinc anode and silver cathode on the parylene substrate. A thin PDMS film is sealed on top of the electrodes for selective detection of oxygen (Fig. 2.8a). The galvanic electrochemical cell generates a voltage around 0.8 V which is enough for the reduction of oxygen. The oxygen level is proportional to the current generated by the galvanic cell (Fig. 2.8b). An integrated electronic system with data readout and wireless transmission is assembled in a compact package using off-the-shelf electronic components (Fig. 2.8c). The sensor and electronics are packed in a 3D printed elastomeric case with

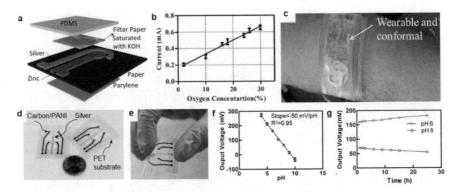

Fig. 2.8 Wound oxygen and pH monitoring using smart bandage. (**a**) Structure of the flexible galvanic oxygen sensor on parylene. (**b**) Calibration plot of the patch oxygen sensor. (**c**) Integrated electronics and the oxygen sensor on a flexible bandage using TangoPlus. (Adapted with permission from Ref. [67]). (**d**) Optical image of flexible pH sensor on PET substrate. (**e**) Optical image showing flexibility of the pH sensor. (**f**) Calibration plot of the pH sensor in range 4–10 with $r^2 = 0.95$ and sensitivity of -50 mV/pH decay. (**g**) Transient response of the pH sensor over 24 h showing its long-term stability. (Adapted with permission from Ref. [71])

exceptional flexibility and tensile strength. The whole device is wearable, conformal to the skin, small, and lightweight.

pH is another important biomarker in the process of wound healing as it is an indicator of wound infection [68, 69]. The pH of a normal healing wound is around 5.5–6.5 but rises to above 6.5 in nonhealing infected wounds [68, 70]. Mostafalu et al. fabricated a pH sensing patch by screen printing carbon/polyaniline (PANI) working electrode and silver reference electrode onto a flexible PET substrate (Fig. 2.8d, e). PANI is a conducting polymer that undergoes protonation and deprotonation in acidic and alkaline environments. The charge accumulation associated with the degree of protonation can be measured as an output voltage signal. Figure 2.8f shows the output voltage is reversely proportional to the pH range of 4–10 with a sensitivity of -50 mV/pH. The signal drift in physiological-relevant pH buffers (pH 6 and 8) is evaluated as shown in Fig. 2.8g. The sensor yields a stable output voltage with less than 6 mV drift over the period of 12 h [71].

2.2.4.2   Smart Bandage for On-demand Drug Delivery

On-demand drug delivery to the wound bed is of crucial importance in wound healing. Traditional wound dressing releases drug in a passive way that is slow and have little spatiotemporal control over the drug release profile. Smart wound dressing is capable of on-demand drug release triggered by internal stimuli such as change in the wound pH, temperature and composition of the extrudate, or external stimuli such as temperature, ultrasound, and electrical and magnetic field [72]. Such platforms need to be flexible to ensure the natural healing process is not impaired from the use of rigid components. In this section, two delivery strategies using temperature and electrical field as external stimuli will be discussed [71, 73, 74].

2.2.4.3 Temperature-Triggered Drug Delivery

Thermal-responsive drug delivery is a widely adopted strategy in smart drug delivery system. The use of thermal-responsive poly(N-isopropylacrylamide) (PNIPAM) has achieved great success in controlled drug delivery due to its suitable lower critical solution temperature (LCST) of ~32 °C, ease of modification, and biocompatibility [75]. In the work reported by Mostafalu et al., drug-laden PNIPAM microparticles are fabricated through microfluidic flow focusing approach. The microparticle undergoes size shrinkage when temperature is increased from 25 °C to 36 °C as shown in Fig. 2.9a, b. To embed the microparticles into the dermal patch, they are loaded into an alginate hydrogel layer using a molding process (Fig. 2.9c). The model drug (fluorescent polymer) is released from the microparticle due to the shrinking effect (Fig. 2.9d). A microheater fabricated by depositing gold on flexible parylene film serves as a thermal actuator for heating the hydrogel layer

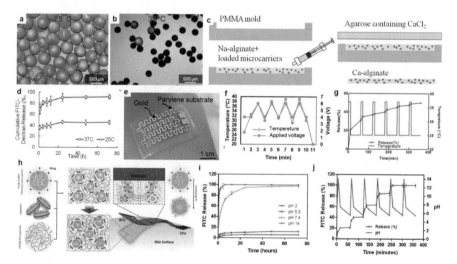

Fig. 2.9 Smart bandage for on-demand drug delivery. (**a** and **b**) Optical images of the thermal-responsive particle at different temperatures. (Adapted with permission from Ref. [71]). (**c**) Fabrication process of the particle-laden hydrogel patch. (**d**) Cumulative release of FITC-dextran from PNIPAM particles at different temperatures. (Adapted with permission from Ref. [73]). (**e**) Optical image of the flexible heater fabricated using gold electrode on parylene substrate. (**f**) Temperature variation in response to cyclic application of voltage to the heater. (**g**) Release rate control of cefazolin by adjusting the temperature. (Adapted with permission from Ref. [71]). (**h**) Schematic illustration of a pH-triggered transdermal drug delivery platform. (**i**) Release profile of FITC from ChPs embedded in PEGDA/laponite hydrogel in different pH. (**j**) Graph showing the control of drug release rate with pH changing. (Adapted with permission from Ref. [74])

(Fig. 2.9e, f). The heater-triggered drug release profile is shown in Fig. 2.9g where the periodically applied heating (30 min on, 30 min off) releases drug (cefazolin) in a dynamic manner.

2.2.4.4   pH-Triggered Drug Delivery

pH-responsive materials can be used for active drug delivery. Kiaee et al. fabricated a drug delivery patch using a pH-sensitive hydrogel containing chitosan particles as drug carriers [74]. In the design, drug-loaded chitosan particles were incorporated into the poly(ethylene glycol)-diacrylate (PEGDA)/laponite hydrogel and coated on the surface of anode electrode (Fig. 2.9h). The applied electrical field causes local pH change near the electrode which disturbs electrostatic interaction between negatively charged chitosan and positively charged laponite. This causes dehydration of chitosan and the subsequent release of encapsulated drugs into the surrounding environment. Removing the applied potential restores original pH and stops drug release. To demonstrate the application of the device, FITC as a model drug was loaded into chitosan particles and its release profile was pH-dependent (Fig. 2.9i). Temporal control of the drug release is shown in Fig. 2.9j where higher pH resulted in greater rate of release. It should be noted that only severe infection of the wound might cause significant pH change, which could only have a beneficial effect on wound through pH-triggered drug release.

2.3   Thread-Based Wearable/Implantable Diagnostics

Textile thread is a highly flexible substrate that offers several intriguing properties for building flexible bioelectronics. Thread can be made out of many natural materials including cotton, silk and wool, or synthetic, man-made materials such as nylon, polyester, and acrylic. The selection of different base material renders thread various properties such as wettability, elasticity, tensile properties, inertness, and conductivity. Thread can be processed using various chemical and physical methods such as dyeing or coating to give it multifunctionality. In addition, thread can be interwoven and used for sewing or suturing, making it an ideal substrate for wearable and implantable diagnostic devices. In this case study, we will discuss the recent research development on thread-based wearable diagnostic therapeutic platforms.

2.3.1   Thread-Based Microfluidics

The passive microfluidic transport in thread is provided by the capillary force. Threads are made from twisted fibers where voids between these fibers form capillary channels that facilitate liquid flow. Some commercially available threads are coated with a layer of wax, which could hinder the fluid transport in them. Removing the wax layer could enhance the wicking property of these threads. On the contrary, the thread can be coated with a water-repellent material to render it hydrophobic. The wettability of commercial cotton threads was controlled by either cleaning with oxygen plasma to render it more hydrophilic or dipping into silicon lubricant to render it more hydrophobic [76]. As shown in Fig. 2.10a, the hydrophilic thread is infused with the green dye with no wicking into the surrounding hydrophobic fabric. As expected, the dye forms droplets on hydrophobic threads with no absorption (Fig. 2.10b). The spatial arrangement of hydrophilic and hydrophobic threads allows for precisely confining the liquid path (Fig. 2.10c). In addition, capillary force does exist even after the initial wetting of the thread. Figure 2.10d shows the subsequent addition of two colorful dyes where the thread still functions as a microfluidic channel after it gets saturated by the first dye. It is well known that capillary flow follows Washburn's equation

$$L = \sqrt{\frac{\gamma r t \cos \emptyset}{2\eta}} \tag{2.1}$$

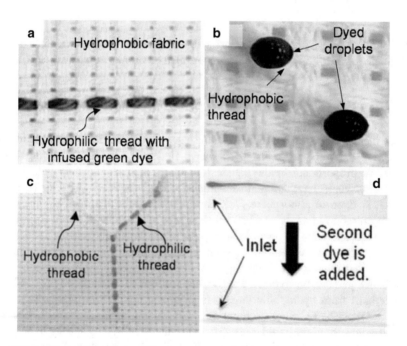

Fig. 2.10 (a) Embroidered hydrophilic thread on a hydrophobic fabric after green dye infusion. (b) Dye droplets on hydrophobic threads. (c) Microfluidic channel formed by hydrophilic and hydrophobic threads. (d) Subsequent addition of two colored dyes on the same thread. (Adapted with permission from Ref. [76])

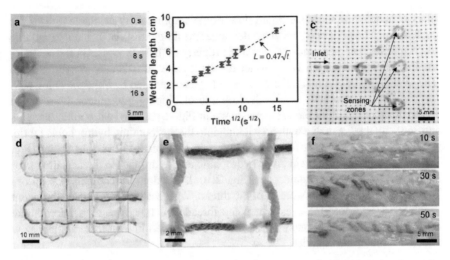

Fig. 2.11 (**a**) Capillary flow of green dye in hydrophilic cotton thread at different time intervals. (**b**) Filling length as a function of the square root of time. (**c**) A three-way splitter microfluidic system made by threads for sensing. (**d**) A 3D microfluidic network by sewing threads onto a PET film. (**e**) Zoomed-in region in (**d**). (**f**) Wicking of blue dye into a thread sutured onto a chicken skin. (Adapted with permission from Ref. [76])

where liquid viscosity (η), surface tension (γ), pore radius (r), and contact angle ($\varnothing$) contribute to the filling length (L). In a controlled environment where η, γ, r, and $\varnothing$ are constant, Eq. (2.1) can be simplified to

$$L = Dt^{1/2} \qquad\qquad (2.2)$$

which means that L is a function of the square root of time. The relationship between L and time is validated as shown in Fig. 2.11a, b. The filling length is linearly proportional to the square root of time. It should be noted that the evaporation could have a significant impact on the capillary behavior of thread when it is open to atmosphere [77]. Evaporation could also lead to changes in the concentration of analytes when using threads as the sampling tool for sensing. However, evaporation is less of an impact when thread is used as implantable devices where the microfluidic system is embedded within the tissue [76]. In the case that continuous sampling over a long period of time is required, an absorbance pad can be connected to the outlet of thread. A three-way microfluidic splitter built from hydrophilic threads is shown in Fig. 2.11c, where the sample (blue dye) can be delivered to three different sensing zone. A three-dimensional microfluidic system for liquid transportation is also achieved by sewing hydrophilic threads in a polyethylene terephthalate (PET) film (Fig. 2.11d, e). Specifically, no sampling mixing is observed during wicking of the threads. The delivery function of threads is demonstrated by sewing the hydrophilic thread onto a chicken skin and wicking the thread from one end (Fig. 2.11f). As shown, no significant leakage

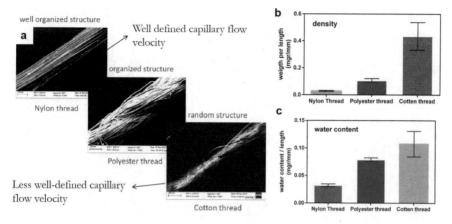

Fig. 2.12 Thread-based microfluidic sampling. (**a**) SEM images of nylon, polyester, and cotton thread. (**b**) Density of three different threads. (**c**) Amount of water taken by three different threads

of sample is observed probably due to the fat tissue surrounding the threads. As mentioned previously, the wicking property of thread makes it a perfect tool for liquid sampling. As a proof of concept, three types of commercially available threads, nylon, polyester, and cotton threads are soaked in water, and their gained weight are measured per unit length (Fig. 2.12). In general, cotton thread samples highest volume of liquid probably due to the high absorbability of cotton fibers. The distinct morphology of different threads as shown by SEM suggests that threads with more organized structure (nylon and polyester) have more precise volume of liquid sampling.

2.3.2   Thread-Based Sensors

In this section, we will highlight the use of thread as sensors for monitoring biophysical and biochemical markers. As most natural and synthetic threads are not conductive, they need to be firstly functionalized with a conductive coating. Figure 2.13 shows SEM images of cotton threads coated by various conductive inks including carbon nanotube, carbon, and conductive polymer. The coating process can be scaled-up using a continuous reel-to-reel dip-coating and drying approach to obtain meter-long conductive threads [76, 78]. Alternatively, electroless deposition has been widely applied to metalize fabrics with high conductivity [79, 80]. We envision this coating method could be applied to threads and is amenable to large-scale fabrication since it is a solution-based approach. Intrinsically conductive threads such as conductive carbon fiber/yarn are also used to build electrochemical sensors due to their good electrical performance [81, 82].

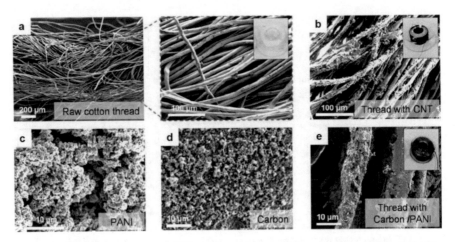

Fig. 2.13 Conductive coating of thread-based sensors. (Adapted with permission from Ref. [76])

Conductive threads can be used directly as strain and pressure sensors. In one example, carbon ink and CNT were coated on elastic polyurethane (PU) threads to make them conductive [76]. A thin layer of PDMS was coated on the threads to protect the conductive layer from scratching and delaminating (Fig. 2.14a). Upon stretching, the resistance of the threads changes linearly with applied tensile strength. The carbon threads can measure strain up to 8% (gauge factor ~ 2), and the CNT threads can measure strain up to 100% (gauge factor ~ 3) (Fig. 2.14b, c). This suggests that the property of ink material could have a significant impact on the sensing performance, as CNT possesses higher deformability due to their fibrous structure [83]. In another example, carbon-coated polybutylene terephthalate (PBT) puffy threads were developed as strain sensors for motion detection [84]. The coiled helical filaments in the PBT threads provide stretchability and display a linear response up to 50% strain (Fig. 2.14d-f). The strain sensing threads are sewn onto a trouser and connected to a microcontroller to record bending of knee as shown in Fig. 2.14g. In addition, the sensing threads are washable after coating a thin layer of PDMS, which does not compromise their sensing capability. In the third example, strain sensing threads are used to realize smart shoe insole for gait monitoring [78]. The threads are arranged in a two-dimensional grid in a highly elastomeric EcoFlex sole for spatial mapping of foot pressure (Fig. 2.14h).

The surface of conductive threads can be modified with bioreceptors as electrochemical sensors. The biorecognition layer could be ion-selective membrane, enzyme, antibody, or aptamer. The immobilization can be achieved through physical absorption, covalent bonding, or electrodeposition. Here we will showcase three relevant works on using thread-based electrochemical sensors for sensing pH, ions, glucose, lactate, and dissolved oxygen. In the work reported by Mostafalu et al., a pH sensing thread is constructed by functionalizing carbon-coated thread with PANI as the working electrode [76]. The sensing thread is connected by a

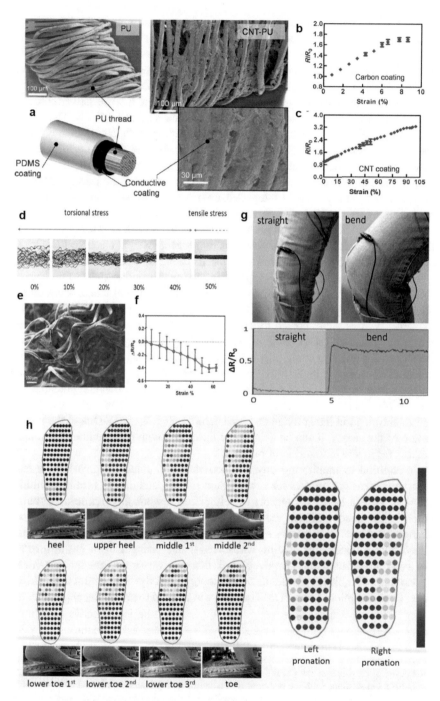

Fig. 2.14 Thread-based physical sensors. (**a**) Sandwich structure of the PU-CNT-PDMS coating and the corresponding SEM images. (**b** and **c**) Variation of the relative resistance as a function of the strain for threads coated with (**b**) carbon ink and (**c**) CNT, respectively. (Adapted with permission

hydrophilic thread that goes into the area of interest for microfluidic sampling (Fig. 2.15a). The thread-based pH sensor has also been validated by stomach and subcutaneous sampling and connected to the wireless transmission electronics (Fig. 2.15a, b). Figure 2.15c, d shows a glucose patch sensor made from embroidered glucose sensing threads. In this case, a glucose oxidase (GOx)-modified carbon thread is used as working electrode; a carbon thread and a silver-/silver chloride–coated thread are used as counter and reference electrodes, respectively. The sensing threads can be readily stitched to any fabric including cloth, and thus they can be used for measuring sweat glucose level. Terse-Thakoor et al. developed a thread-based multiplexed sensor patch for real-time sweat monitoring [85]. In this work, an array of thread sensors for electrolytes (sodium and ammonium ions), metabolites (lactate), and acidity (pH) is integrated onto a bandage patch for on-skin sweat measurement (Fig. 2.15e). The performance of individual thread sensor is shown in Fig. 2.15f, where the ion sensor, pH sensor, and lactate sensor all display satisfactory linear range, limit of detection, and sensitivity. Moreover, the sensor is validated in vivo by placing the patch sensor on the forehead of the human subject during a maximal exertion test (Fig. 2.15g). It is also worth noting that a miniaturized circuit module containing a potentiostat, microprocessor, and wireless circuitry is connected to the patch sensor for wireless smartphone readout [85]. In the last example, flexible silver-coated threads are explored as gas sensors for measuring tissue oxygenation (pO_2) [86]. A "wire-type" and "tip-type" oxygen sensor is developed for either probing pO_2 over a large surface area or pin-point local pO_2 at region of interest (Fig. 2.15h). Two silver-coated threads are used as cathode and anode, where oxygen is reduced at the cathode at a given polarization potential, and the resulting current is measured as a function of pO_2. The thread-based oxygen sensor shows good linear range and repeatability (Fig. 2.15i, j). Due to the flexible nature of the sensor, it can be potentially integrated with tissue-embedded sensing devices for in situ monitoring of oxygen.

In addition to monitoring physical activities and soluble biomarkers, threads were utilized as gas sensors for environment sensing and human health monitoring [87–90]. Thread-based gas sensors use either colorimetric or electronic readout. In colorimetric readout, threads are coated by gas-sensitive dyes which undergo color change in contact with the gas species. In electronic readout, threads are functionalized with a conductive coating whose resistance changes upon gas adsorption. As a direct measurement, Owyeung et al. fabricated wearable gas sensing threads by loading pH indicator dyes and a metalloporphyrin dye into cotton threads [88]. After dye adsorption, a thin PDMS coating was applied to physically entrap the dyes

---

Fig. 2.14 (continued) from Ref. [76]). (**d**) Entanglement of the C-PBT thread filament up 50% strain followed by tensile stress after 50% strain. (**e**) SEM image of the C-PBT thread. (**f**) Piezoresistive response of the C-PBT sensor to strain. (**g**) Demonstration of the stitch-ability of the C-PBT sensor along with the response of the sensor in application. (Adapted with permission from Ref. [84]). (**h**) Thread-based strain sensor for gait analysis. (Adapted with permission from Ref. [78])

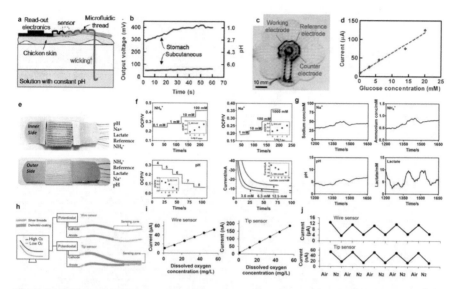

Fig. 2.15 Thread-based electrochemical sensors. (**a**) Schematic showing a thread-based pH sensor for measuring oH in a vitro model. (**b**) pH reading of the sensor in the stomach and under the skin from a living mouse. (**c**) Optical image of a thread-based glucose sensor integrated into a patch. (**d**) Calibration plot of the glucose sensor. (**e**) Image of a thread-based sweat patch sensor. (**f**) Calibration plot of the sweat patch sensor for measuring ammonia, sodium, pH, and lactate. (**g**) Real-time measurement of the abovementioned analyte from a human subject during exercise. (**h**) Schematic illustration of a thread-based "tip" and "wire" type oxygen sensor. (**i**) Calibration plot of the "wire" and "tip" oxygen sensor. (**j**) Repeatability test of the thread-based oxygen sensor

in order to prevent dye leaching when used in aqueous environment. As a proof of concept, the sensing threads were tested for ammonia and hydrogen chloride, with a range of detection of 50–1000 ppm. A smartphone app was developed for colorimetric readout which can extract red (R), green (G), and blue (B) channels of the acquired images to detect presence of an analyte. The gas sensing threads can be readily sown into clothing as wearable gas monitors (Fig. 2.16).

2.3.3 Thread-Based Drug Delivery

Threads are multifilament fibers with large surface-to-volume ratio which are advantageous as drug carriers. There has been a long history using surgical sutures to deliver anti-bacterial reagents for anti-infection and recently growth factors to promote wound healing at the wound site [91, 92]. In most of these studies, passive and sustained drug release is achieved by gradual dissolution of the polymer matrixes such as PGA and PLGA, and they lack both spatial and temporal control of the release profile. On the other hand, on-demand drug delivery triggered by an stimuli is favorable as the drug dosage can be well controlled to enhance

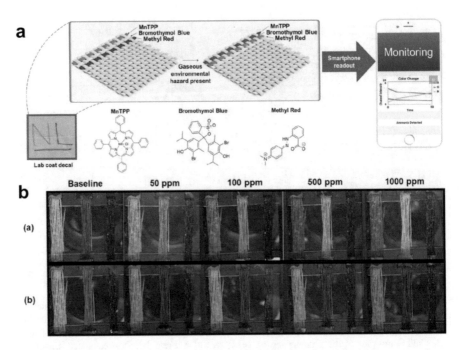

Fig. 2.16 Thread-based gas sensing. (**a**) Schematics illustrating thread-based gas sensors integrated in a textile patch. (**b**) Optical images of thread-based gas sensor exposed to different concentrations of ammonia or hydrochloride acid

therapeutical efficacy. Methods to trigger drug release include internal stimuli such as pH and temperature change at the wound bed; or external stimuli such as temperature, ultrasound, and electrical and magnetic field.

A thread-based controlled drug delivery platform for wound healing was developed by Mostafalu et al. with enhanced spatiotemporal resolution [93]. In this work, composite threads consisted of an inner conductive core and an outer drug-loaded hydrogel layer were assembled into a textile patch using textile processing. The temperature-triggered drug release was realized by encapsulating drugs into thermal-responsive microparticles embedded in the hydrogel layer as mentioned previously in "smart-bandage for on-demand drug delivery." Each of these microheaters functioned as an operating unit and can be individually triggered by a microcontroller to realize spatial control of drug release (Fig. 2.17a). To demonstrate the efficacy of the platform, two types of therapeutical reagents, antibiotics, and vascular endothelial growth factor (VEGF) were loaded into threads and cultured with bacterial-infected cells using a co-culture system (Fig. 2.17b). The results showed bacteria were eradicated at 4 h due to release of antibiotics and the formation of capillaries at 8 h (and indication of angiogenesis) due to release of VEGF.

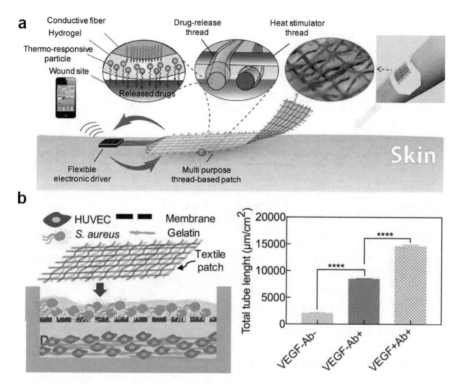

Fig. 2.17 Thread-based drug delivery. (**a**) Schematic of a multipurpose thread-based patch for transdermal drug delivery with a hydrogel layer carrying thermal-responsive particle coated on a flexible thread-based heater. (**b**) Left: In vitro model for co-culturing of HUVECs and *Staphylococcus aureus* bacteria treated with the patch loaded with VEGF and antibiotic coated fiber. Right: Quantitative analysis of total tube length formed when treated with different conditions. (Adapted with permission from Ref. [93])

2.3.4   Thread-Based Transistor and Circuits

The realization of thread-based transistor (TBT) is intriguing as transistors are one of the basic electronic components in any circuit. As the number of available sensors increases, the integration of electronics for multiplexed readout and amplification is needed. Conventional silicon-based transistors and microelectronics are unsuited because of the mismatch between the rigid, inflexible silicon-based devices, and the soft, flexible biological tissues. The development of TBT can enable an all-thread multiplexed diagnostic platform by interconnection with sampling and sensing threads.

The basic idea for constructing thread-based transistor is shown in Fig. 2.18a. Two threads, functioning as gate and drain-source, respectively, are connected by

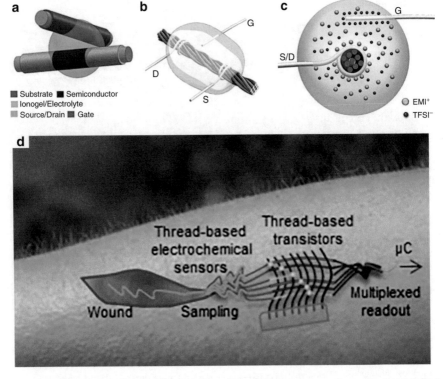

Fig. 2.18 Thread-based transistors. (**a**) Schematic illustration of 3D transistor geometry realized on thread/fibers. (**b**) Design of a thread-based transistor using ionic liquid gel. Gold wires are used as gate, drain/source electrodes. (**c**) Side view of the geometry from (**b**) that highlights the thread-based transistor geometry with ionogel gate electrolyte enabling all-around electrostatic gating. (**d**) Schematic illustration of an all-thread-based multiplexed diagnostic platform. (Adapted with permission from Ref. [96])

ionogel or electrolyte as gate dielectric. The advantage of utilizing 3D thread/fiber substrates compared with traditional 2D silicon-based transistor is the much larger surface-to-volume ratio which could enhance device performance and sensitivity. The idea of electrolyte gating also solved issues of uneven deposition of dielectric materials onto 3D thread/fiber substrates which could negatively impact the device performance [94, 95].

Sonkusale and his group has reported a type of TBT that uses ionogel dielectric for electrolyte gating [96]. The geometry of the device is shown in Fig. 2.18b, c. A linen thread coated by semiconducting material functions as the active channel where two gold wires were knotted to the sides as drain and source electrodes. Ionogel solution was pasted on top and a third gold wire was placed on top of the gel as gate electrode. The electrostatic double layer (EDL) capacitance formed at the interface of electrolytes and semiconductor/metal electrodes provides basic operation of the transistor, which is largely unaffected by the thickness of the dielectric material. This is advantageous for making transistors on rough surfaces without the need for pre-deposition of a polymer layer. In addition, the use of ionogel provides a large capacitance and a low operation voltage. We further integrated TBT with thread-based electrochemical sensors for multiplexed diagnostics, where sodium and ammonium sensing were achieved using separate working electrodes paired with TBT at discrete locations [96]. The conceptual illustration of an all-thread-based platform for wound monitoring is shown in Fig. 2.18d. The sampling threads sample wound extrudes through wicking effect and deliver them to the sensing threads. Individual sensing thread is selected electronically for readout by TBTs which serve as an electronic switch and was configured as a multiplexer. Recently, we increased the throughput of fabrication of thread-based transistors using a stencil-based patterning approach that defined the channel gap for number of transistors.

2.4 Challenges and Outlook

This chapter provided an overview of flexible sensors, microfluidics, and drug delivery. This chapter began with a background on various kinds of biophysical and biochemical signals of interest and ended with select case studies of flexible bioelectronic platforms such as smart bandages. To achieve flexibility, devices were fabricated primarily using solution-based processing such as dip coating or screen printing of underlying flexible substrates such as thread, paper, or other polymers. Sensing functionality in such devices is achieved using appropriate ink formulation. In all these applications, there is still a need to power them up either using battery or energy harvesting from the environment. While this was not adequately discussed, flexible realizations of battery and supercapacitors are expected to be a topic of important consideration [97, 98]. In this chapter, our primary focus was on achieving flexibility; however, other metrics such as elasticity and stretchability may be more important [56–58]. For example, in strain sensors to monitor physical motion, underlying sensors are expected to be stretchable [78, 84]. Going forward, with increased tissue integration, stretchable and even elastic bioelectronics may become more of a topic of research interest. Finally, the materials used in the realizations of these flexible bioelectronic platforms currently range from synthetic polymers to natural biopolymers. Fouling of electrodes and materials when in contact with biological fluid is an important concern. With increased tissue and body integration, sensors and electronics in the future will be realized using biopolymers and naturally derived materials to avoid foreign body response and for enhanced biocompatibility.

References

1. Y. Yang, X. Yang, Y. Tan, Q. Yuan, Recent progress in flexible and wearable bio-electronics based on nanomaterials. Nano Res. **10**, 1560–1583 (2017)
2. C. Liao, M. Zhang, M.Y. Yao, T. Hua, L. Li, F. Yan, Flexible organic electronics in biology: Materials and devices. Adv. Mater. **27**, 7493–7527 (2015)
3. Wearable Sensors Market Report: Trends, Forecast and Competitive Analysis. Available online   http://www.researchandmarkets.com/reports/5002980/wearable-sensors-market-report-trends-forecast?utm_source=GNOM&utm_medium=PressRelease&utm_code=n6j8b4&utm_campaign=1470719+-+World+Wearable+Sensors+Market+Report+2021+-+Market+is+Expected+to+Grow+with+a+CAGR+of+30%25+from+2019+to+2024&utm_exec=chdo54prd. Accessed on 27 Jan 20121 (accessed 27 June 2019)
4. G. Acar, O. Ozturk, A.J. Golparvar, T.A. Elboshra, K. Böhringer, M.K. Yapici, Wearable and flexible textile electrodes for biopotential signal monitoring: A review. Electronics **8** (2019)
5. M. Cuartero, M. Parrilla, G.A. Crespo, Wearable potentiometric sensors for medical applications. Sensors (Basel) **19**, 363 (2019)
6. Z.S. Ballard, A. Ozcan, Wearable Optical Sensors, *Mobile Health*, (2017), pp. 313–342
7. M.A. Arnold, M.E. Meyerhoff, Ion-selective electrodes. Anal. Chem. **56**, 20–48 (2002)
8. G. Dimeski, T. Badrick, A.S. John, Ion Selective Electrodes (ISEs) and interferences–a review. Clin. Chim. Acta **411**, 309–317 (2010)
9. C. Chen, Q. Xie, D. Yang, H. Xiao, Y. Fu, Y. Tan, et al., Recent advances in electrochemical glucose biosensors: a review. RSC Adv. **3**, 4473 (2013)
10. K. Ramanathan, M. Rank, J. Svitel, A. Dzgoev, B. Danielsson, The development and applications of thermal biosensors for bioprocess monitoring. Trends Biotechnol. **17**, 499–505 (1999)
11. B. Zuo, S. Li, Z. Guo, J. Zhang, C. Chen, Piezoelectric Immunosensor for SARS-associated coronavirus in sputum. Anal. Chem. **76**, 3536–3540 (2004)
12. R. Chauhan, J. Singh, P.R. Solanki, T. Manaka, M. Iwamoto, T. Basu, et al., Label-free piezo-electric immunosensor decorated with gold nanoparticles: Kinetic analysis and biosensing application. Sensors Actuators B Chem. **222**, 804–814 (2016)
13. A. Rasheed, E. Iranmanesh, W. Li, H. Ou, A. S. Andrenko, K. Wang. A wearable autonomous heart rate sensor based on piezoelectric-charge-gated thin-film transistor for continuous multi-point monitoring, *2017 39th Annual International Conference of the IEEE Engineering in Medicine and Biology Society (EMBC)* (2017)
14. S.Y. Chung, H.-J. Lee, T.I. Lee, Y.S. Kim, A wearable piezoelectric bending motion sensor for simultaneous detection of bending curvature and speed. RSC Adv. **7**, 2520–2526 (2017)
15. L. Mccullough, S. Arora, Diagnosis and treatment of hypothermia. Am. Fam. Physician **70**, 2325–2332 (2004)
16. H. Pang, Y.-C. Zhang, T. Chen, B.-Q. Zeng, Z.-M. Li, Tunable positive temperature coefficient of resistivity in an electrically conducting polymer/graphene composite. Appl. Phys. Lett. 96, 251907 (2010)
17. W.P. Shih, L.C. Tsao, C.W. Lee, M.Y. Cheng, C. Chang, Y.J. Yang, et al., Flexible temperature sensor array based on a graphite-polydimethylsiloxane composite. Sensors (Basel) **10**, 3597–3610 (2010)
18. T. Yokota, Y. Inoue, Y. Terakawa, J. Reeder, M. Kaltenbrunner, T. Ware, et al., Ultraflexible, large-area, physiological temperature sensors for multipoint measurements. Proc. Natl. Acad. Sci. U. S. A. **112**, 14533–14538 (2015)
19. Y. Chen, B. Lu, Y. Chen, X. Feng, Breathable and Stretchable Temperature Sensors Inspired by Skin. Sci. Rep. **5**, 11505 (2015)
20. J. Yang, D. Wei, L. Tang, X. Song, W. Luo, J. Chu, et al., Wearable temperature sensor based on graphene nanowalls. RSC Adv. **5**, 25609–25615 (2015)

21. T.I. Gonzales, J.Y. Jeon, T. Lindsay, K. Westgate, I. Perez-Pozuelo, S. Hollidge, et al., Resting heart rate as a biomarker for tracking change in cardiorespiratory fitness of UK adults: The Fenland Study. medRxiv (2020). https://doi.org/10.1101/2020.07.01.20144154

22. M. Saad, L.B. Ray, B. Bujaki, A. Parvaresh, I. Palamarchuk, J. De Koninck, et al., Using heart rate profiles during sleep as a biomarker of depression. BMC Psychiatr. **19**, 168 (2019)

23. U. Rajendra Acharya, K. Paul Joseph, N. Kannathal, C.M. Lim, J.S. Suri, Heart rate variability: A review. Med. Biol. Eng. Comput. **44**, 1031–1051 (2006)

24. T. Sekine, R. Sugano, T. Tashiro, J. Sato, Y. Takeda, H. Matsui, et al., Fully Printed Wearable Vital Sensor for Human Pulse Rate Monitoring using Ferroelectric Polymer. Sci. Rep. **8**, 4442 (2018)

25. S. Yoon, J.K. Sim, Y.H. Cho, A flexible and wearable human stress monitoring patch. Sci. Rep. **6**, 23468 (2016)

26. M. Grassmann, E. Vlemincx, A. von Leupoldt, J.M. Mittelstadt, O. Van den Bergh, Respiratory changes in response to cognitive load: A systematic review. Neural Plast. **2016**, 8146809 (2016)

27. M.A. Cretikos, R. Bellomo, K. Hillman, J. Chen, S. Finfer, A. Flabouris, Respiratory rate: The neglected vital sign. Med. J. Aust. **188**, 657–659 (2008)

28. M. Ciocchetti, C. Massaroni, P. Saccomandi, M.A. Caponero, A. Polimadei, D. Formica, et al., Smart Textile Based on Fiber Bragg Grating Sensors for Respiratory Monitoring: Design and Preliminary Trials. Biosensors (Basel) **5**, 602–615 (2015)

29. B. Li, G. Xiao, F. Liu, Y. Qiao, C.M. Li, Z. Lu, A flexible humidity sensor based on silk fabrics for human respiration monitoring. J. Mater. Chem. C **6**, 4549–4554 (2018)

30. S.W. Park, P.S. Das, A. Chhetry, J.Y. Park, A flexible capacitive pressure sensor for wearable respiration monitoring system. IEEE Sensors J. **17**, 1 (2017)

31. N. Luo, W. Dai, C. Li, Z. Zhou, L. Lu, C.C.Y. Poon, et al., Flexible Piezoresistive sensor patch enabling ultralow power Cuffless blood pressure measurement. Adv. Funct. Mater. **26**, 1178–1187 (2016)

32. S. Xu, Y. Zhang, J. Cho, J. Lee, X. Huang, L. Jia, et al., Stretchable batteries with self-similar serpentine interconnects and integrated wireless recharging systems. Nat. Commun. **4**, 1543 (2013)

33. J.-Y. Baek, J.-H. An, J.-M. Choi, K.-S. Park, S.-H. Lee, Flexible polymeric dry electrodes for the long-term monitoring of ECG. Sensors Actuators A Phys. **143**, 423–429 (2008)

34. L.-F. Wang, J.-Q. Liu, B. Yang, C.-S. Yang, PDMS-based low cost flexible dry electrode for long-term EEG measurement. IEEE Sensors J. **12**, 2898–2904 (2012)

35. H.C. Jung, J.H. Moon, D.H. Baek, J.H. Lee, Y.Y. Choi, J.S. Hong, et al., CNT/PDMS composite flexible dry electrodes for long-term ECG monitoring. I.E.E.E. Trans. Biomed. Eng. **59**, 1472–1479 (2012)

36. Y. Yamamoto, D. Yamamoto, M. Takada, H. Naito, T. Arie, S. Akita, et al., Efficient skin temperature sensor and stable gel-less sticky ECG sensor for a wearable flexible healthcare patch. Adv. Healthc. Mater. **6** (2017)

37. A. Miyamoto, S. Lee, N.F. Cooray, S. Lee, M. Mori, N. Matsuhisa, et al., Inflammation-free, gas-permeable, lightweight, stretchable on-skin electronics with nanomeshes. Nat. Nanotechnol. **12**, 907–913 (2017)

38. Y.J. Wang, C.H. Chen, C.Y. Sue, W.H. Lu, Y.H. Chiou, Estimation of Blood Pressure in the Radial Artery Using Strain-Based Pulse Wave and Photoplethysmography Sensors. Micromachines (Basel) **9** (2018)

39. Q. Yu, L. Xue, J. Hiblot, R. Griss, S. Fabritz, C. Roux, et al., Semisynthetic sensor proteins enable metabolic assays at the point of care. Science **361**, 1122–1126 (2018)

40. W.U. Dittmer, T.H. Evers, W.M. Hardeman, W. Huijnen, R. Kamps, P. de Kievit, et al., Rapid, high sensitivity, point-of-care test for cardiac troponin based on optomagnetic biosensor. Clin. Chim. Acta **411**, 868–873 (2010)

41. Y. Song, Y.Y. Huang, X. Liu, X. Zhang, M. Ferrari, L. Qin, Point-of-care technologies for molecular diagnostics using a drop of blood. Trends Biotechnol. **32**, 132–139 (2014)

42. Y. Yu, I. Prassas, C.M. Muytjens, E.P. Diamandis, Proteomic and peptidomic analysis of human sweat with emphasis on proteolysis. J. Proteome **155**, 40–48 (2017)

43. S. Jadoon, S. Karim, M.R. Akram, A. Kalsoom Khan, M.A. Zia, A.R. Siddiqi, et al., Recent developments in sweat analysis and its applications. Int. J. Anal. Chem. **2015**, 164974 (2015)

44. A. Koh, D. Kang, Y. Xue, S. Lee, R.M. Pielak, J. Kim, et al., A soft, wearable microfluidic device for the capture, storage, and colorimetric sensing of sweat. Sci. Transl. Med. **8**, 366ra165 (2016)

45. W. Gao, S. Emaminejad, H.Y.Y. Nyein, S. Challa, K. Chen, A. Peck, et al., Fully integrated wearable sensor arrays for multiplexed in situ perspiration analysis. Nature **529**, 509–514 (2016)

46. P.P. Samant, M.M. Niedzwiecki, N. Raviele, V. Tran, J. Mena-Lapaix, D.I. Walker, et al., Sampling interstitial fluid from human skin using a microneedle patch. Sci. Transl. Med. **12**, eaaw0285 (2020)

47. P.R. Miller, R.M. Taylor, B.Q. Tran, G. Boyd, T. Glaros, V.H. Chavez, et al., Extraction and biomolecular analysis of dermal interstitial fluid collected with hollow microneedles. Commun. Biol. **1**, 173 (2018)

48. P.P. Samant, M.R. Prausnitz, Mechanisms of sampling interstitial fluid from skin using a microneedle patch. Proc. Natl. Acad. Sci. U. S. A. **115**, 4583–4588 (2018)

49. J. Madden, C. O'Mahony, M. Thompson, A. O'Riordan, P. Galvin, Biosensing in dermal interstitial fluid using microneedle based electrochemical devices. Sens. Bio. Sens. Res. **29**, 100348 (2020)

50. A.J. Bandodkar, W. Jia, C. Yardimci, X. Wang, J. Ramirez, J. Wang, Tattoo-based noninvasive glucose monitoring: A proof-of-concept study. Anal. Chem. **87**, 394–398 (2015)

51. D. Malamud, Saliva as a diagnostic fluid. Dent. Clin. N. Am. **55**, 159–178 (2011)

52. K. Aro, F. Wei, D.T. Wong, M. Tu, Saliva liquid biopsy for point-of-care applications. Front. Public Health **5**, 77 (2017)

53. J. Kim, S. Imani, W.R. de Araujo, J. Warchall, G. Valdes-Ramirez, T.R. Paixao, et al., Wearable salivary uric acid mouthguard biosensor with integrated wireless electronics. Biosens. Bioelectron. **74**, 1061–1068 (2015)

54. P. Mochalski, G. Shuster, M. Leja, K. Unterkofler, C. Jaeschke, R. Skapars, et al., Non-contact breath sampling for sensor-based breath analysis. J. Breath Res. **13**, 036001 (2019)

55. A. Gholizadeh, D. Voiry, C. Weisel, A. Gow, R. Laumbach, H. Kipen, et al., Toward point-of-care management of chronic respiratory conditions: Electrochemical sensing of nitrite content in exhaled breath condensate using reduced graphene oxide. Microsyst. Nanoeng. **3**, 17022 (2017)

56. H.U. Chung, B.H. Kim, J.Y. Lee, J. Lee, Z. Xie, E.M. Ibler, et al., Binodal, wireless epidermal electronic systems with in-sensor analytics for neonatal intensive care. Science **363**, 6430 (2019)

57. A. Chortos, G.I. Koleilat, R. Pfattner, D. Kong, P. Lin, R. Nur, et al., Mechanically durable and highly stretchable transistors employing carbon nanotube semiconductor and electrodes. Adv. Mater. **28**, 4441–4448 (2016)

58. A. Chortos, J. Lim, J. W. To, M. Vosgueritchian, T.J. Dusseault, T.H. Kim, et al., Highly stretchable transistors using a microcracked organic semiconductor. Adv. Mater. **26**, 4253–4259 (2014)

59. J. Liang, L. Li, D. Chen, T. Hajagos, Z. Ren, S.Y. Chou, et al., Intrinsically stretchable and transparent thin-film transistors based on printable silver nanowires, carbon nanotubes and an elastomeric dielectric. Nat. Commun. **6**, 7647 (2015)

60. S. Wang, J. Xu, W. Wang, G.N. Wang, R. Rastak, F. Molina-Lopez, et al., Skin electronics from scalable fabrication of an intrinsically stretchable transistor array. Nature **555**, 83–88 (2018)

61. T.N. Demidova-Rice, M.R. Hamblin, I.M. Herman, Acute and impaired wound healing: Pathophysiology and current methods for drug delivery, part 1: Normal and chronic wounds: Biology, causes, and approaches to care. Adv. Skin Wound Care **25**, 304–314 (2012)

62. R.G. Frykberg, J. Banks, Challenges in the Treatment of Chronic Wounds. Adv. Wound Care (New Rochelle) **4**, 560–582 (2015)

63. W.H. Eaglstein, V. Falanga, Chronic wounds. Surg. Clin. North Am. **77**, 689–700 (1997)
64. S.R. Nussbaum, M.J. Carter, C.E. Fife, J. DaVanzo, R. Haught, M. Nusgart, et al., An economic evaluation of the impact, cost, and medicare policy implications of chronic nonhealing wounds. Value Health **21**, 27–32 (2018)
65. F. Gottrup, Oxygen in wound healing and infection. World J. Surg. **28**, 312–315 (2004)
66. S. Schreml, R.M. Szeimies, L. Prantl, S. Karrer, M. Landthaler, P. Babilas, Oxygen in acute and chronic wound healing. Br. J. Dermatol. **163**, 257–268 (2010)
67. P. Mostafalu, W. Lenk, M.R. Dokmeci, B. Ziaie, A. Khademhosseini, S.R. Sonkusale, Wireless flexible smart bandage for continuous monitoring of wound oxygenation. IEEE Trans. Biomed. Circuits Syst. **9**, 670–677 (2015)
68. L.A. Schneider, A. Korber, S. Grabbe, J. Dissemond, Influence of pH on wound-healing: A new perspective for wound-therapy? Arch. Dermatol. Res. **298**, 413–420 (2007)
69. S.L. Percival, S. McCarty, J.A. Hunt, E.J. Woods, The effects of pH on wound healing, biofilms, and antimicrobial efficacy. Wound Repair Regen. **22**, 174–186 (2014)
70. R. Rahimi, M. Ochoa, A. Tamayol, S. Khalili, A. Khademhosseini, B. Ziaie, Highly stretchable potentiometric pH sensor fabricated via laser carbonization and machining of carbon-polyaniline composite. ACS Appl. Mater. Interfaces **9**, 9015–9023 (2017)
71. P. Mostafalu, A. Tamayol, R. Rahimi, M. Ochoa, A. Khalilpour, G. Kiaee, et al., Smart bandage for monitoring and treatment of chronic wounds. Small, e1703509 (2018)
72. F. Meng, Z. Zhong, J. Feijen, Stimuli-responsive polymersomes for programmed drug delivery. Biomacromolecules **10**, 197–209 (2009)
73. S. Bagherifard, A. Tamayol, P. Mostafalu, M. Akbari, M. Comotto, N. Annabi, et al., Dermal patch with integrated flexible heater for on demand drug delivery. Adv. Healthc. Mater. **5**, 175–184 (2016)
74. G. Kiaee, P. Mostafalu, M. Samandari, S. Sonkusale, A pH-mediated electronic wound dressing for controlled drug delivery. Adv. Healthc. Mater. **7**, e1800396 (2018)
75. Y. Guan, Y. Zhang, PNIPAM microgels for biomedical applications: From dispersed particles to 3D assemblies. Soft Matter **7**, 6375 (2011)
76. P. Mostafalu, M. Akbari, K.A. Alberti, Q. Xu, A. Khademhosseini, S.R. Sonkusale, A toolkit of thread-based microfluidics, sensors, and electronics for 3D tissue embedding for medical diagnostics. Microsyst. Nanoeng. **2**, 16039 (2016)
77. R. Safavieh, G.Z. Zhou, D. Juncker, Microfluidics made of yarns and knots: From fundamental properties to simple networks and operations. Lab Chip **11**, 2618–2624 (2011)
78. F. Alaimo, A. Sadeqi, H. Rezaei Nejad, Y. Jiang, W. Wang, D. Demarchi, et al., Reel-to-reel fabrication of strain sensing threads and realization of smart insole. Sens. Actuators. A **301**, 111741 (2020)
79. M. Montazer, V. Allahyarzadeh, Electroless plating of silver nanoparticles/nanolayer on polyester fabric using AgNO3/NaOH and ammonia. Ind. Eng. Chem. Res. **52**, 8436–8444 (2013)
80. S. Mu, H. Xie, W. Wang, D. Yu, Electroless silver plating on PET fabric initiated by in situ reduction of polyaniline. Appl. Surf. Sci. **353**, 608–614 (2015)
81. M. Sekar, M. Pandiaraj, S. Bhansali, N. Ponpandian, C. Viswanathan, Carbon fiber based electrochemical sensor for sweat cortisol measurement. Sci. Rep. **9**, 403 (2019)
82. M. Parrilla, J. Ferré, T. Guinovart, F.J. Andrade, Wearable potentiometric sensors based on commercial carbon fibres for monitoring sodium in sweat. Electroanalysis **28**, 1267–1275 (2016)
83. T. Yamada, Y. Hayamizu, Y. Yamamoto, Y. Yomogida, A. Izadi-Najafabadi, D.N. Futaba, et al., A stretchable carbon nanotube strain sensor for human-motion detection. Nat. Nanotechnol. **6**, 296–301 (2011)
84. A. Sadeqi, H.R. Nejad, F. Alaimo, H. Yun, M. Punjiya, S. Sonkusale, Washable smart threads for strain sensing fabrics. IEEE Sensors J., 1 (2018)
85. T. Terse-Thakoor, M. Punjiya, Z. Matharu, B. Lyu, M. Ahmad, G.E. Giles, et al., Thread-based multiplexed sensor patch for real-time sweat monitoring. npj Flexib. Electron. **4**, 18 (2020)

86. J. Xia, S. Sonkusale, Flexible thread-based electrochemical sensors for oxygen monitoring. Analyst **146**, 2983–2990 (2021)

87. D. H. Kim, J. H. Cha, J. Y. Lim, J. Bae, W. Lee, K. R. Yoon, et al., Colorimetric dye-loaded nanofiber yarn: eye-readable and weavable gas sensing platform. ACS Nano, **14**(12), 16907–16918 (2020)

88. R.E. Owyeung, M.J. Panzer, S.R. Sonkusale, Colorimetric gas sensing washable threads for smart textiles. Sci. Rep. **9**, 5607 (2019)

89. N. Indarit, Y.-H. Kim, N. Petchsang, R. Jaisutti, Highly sensitive polyaniline-coated fiber gas sensors for real-time monitoring of ammonia gas. RSC Adv. **9**, 26773–26779 (2019)

90. T. Seesaard, P. Lorwongtragool, T. Kerdcharoen, Development of fabric-based chemical gas sensors for use as wearable electronic noses. Sensors (Basel) **15**, 1885–1902 (2015)

91. S. Rothenburger, D. Spangler, S. Bhende, D. Burkley, In vitro antimicrobial evaluation of coated VICRYL* plus antibacterial suture (coated polyglactin 910 with triclosan) using zone of inhibition assays. Surg. Infect. **3**(Suppl 1), S79–S87 (2002)

92. B. Joseph, A. George, S. Gopi, N. Kalarikkal, S. Thomas, Polymer sutures for simultaneous wound healing and drug delivery – a review. Int. J. Pharm. **524**, 454–466 (2017)

93. P. Mostafalu, G. Kiaee, G. Giatsidis, A. Khalilpour, M. Nabavinia, M.R. Dokmeci, et al., A textile dressing for temporal and dosage controlled drug delivery. Adv. Funct. Mater. **27**, 1702399 (2017)

94. M. Hamedi, R. Forchheimer, O. Inganas, Towards woven logic from organic electronic fibres. Nat. Mater. **6**, 357–362 (2007)

95. M. Hamedi, L. Herlogsson, X. Crispin, R. Marcilla, M. Berggren, O. Inganas, Fiber-embedded electrolyte-gated field-effect transistors for e-textiles. Adv. Mater. **21**, 573–577 (2009)

96. R.E. Owyeung, T. Terse-Thakoor, H. Rezaei Nejad, M.J. Panzer, S.R. Sonkusale, Highly flexible transistor threads for all-thread based integrated circuits and multiplexed diagnostics. ACS Appl. Mater. Interfaces **11**, 31096–31104 (2019)

97. Y. Yu, J. Nassar, C. Xu, J. Min, Y. Yang, A. Dai, et al., Biofuel-powered soft electronic skin with multiplexed and wireless sensing for human-machine interfaces. Sci. Robot. **5**, eaaz7946 (2020)

98. Z. Su, H. Wu, H. Chen, H. Guo, X. Cheng, Y. Song, et al., Digitalized self-powered strain gauge for static and dynamic measurement. Nano Energy **42**, 129–137 (2017)

Chapter 3
Low-Noise CMOS Signal Conditioning Circuits

3.1 Introduction to Discrete and CMOS Signal Conditioning

Flexible sensor systems are used in virtually all aspects of our day-to-day life, including monitoring, security, surveillance, and awareness in general. In addition, sensors are widely used in industrial and scientific applications to monitor physical and chemical parameters such as temperature, pressure, displacement, acceleration, force, gas concentration, droplet detection, and humidity. Moreover, sensors are widely used in wearable and implantable systems for monitoring physiological signals [1–5].

The block diagram of a flexible sensors system is shown in Fig. 3.1. Flexible sensors represent the first and one of the most important elements in the measurement and instrumentation systems. It converts the quantity to be sensed (measurand) into a quantity—either change in voltage, resistance, capacitance, and so on based on the sensor type—suitable for the signal conditioning circuit.

Depending on the type of flexible sensor, the sensor's signal may or may not be compatible with the input requirement of the data acquisition system. In addition, the signal generated by the sensor is often too weak and noisy. Moreover, it may contain undesirable components. A signal conditioning circuit's function is to convert the sensor signal into a format compatible with the data acquisition unit. It can have one or more operations, such as filtering, amplification, and linearization.

The flexible sensors can be categorized broadly into two categories: passive and active sensors. Passive sensors are those who do not require any external power to convert the measurand into an equivalent electrical signal. Such flexible sensors provide a direct electrical signal in response to the quantity to be sensed. For example, a thermocouple, a piezoelectric sensor, and a photodiode are the passive

Dr. Maryam Shojaei Baghini (IIT Bombay), Dr. Meraj Ahmad (IIT Bombay) and Dr. Shahid Mailk (IIT Delhi) contributed to this chapter.

© The Author(s), under exclusive license to Springer Nature Switzerland AG 2022
S. Sonkusale et al., *Flexible Bioelectronics with Power Autonomous Sensing and Data Analytics*, https://doi.org/10.1007/978-3-030-98538-7_3

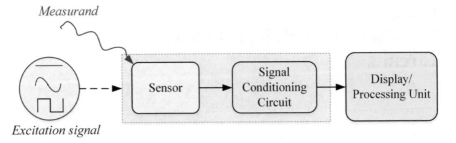

Measurand

Excitation signal

Fig. 3.1 Block diagram of a flexible sensors system

sensors. On the other hand, active sensors need the power source for converting the sensor information into the equivalent electrical form. The resistive and capacitive sensors are examples of active sensors [6].

3.1.1 Sensors

Resistive and capacitive sensors are the most commonly used sensors in industrial, scientific, and medical applications. These sensors convert the measurand, which can be either a physical, chemical, or a biological quantity, into a change in either the resistance or capacitance based on the sensor's topology.

3.1.1.1 Capacitive Sensors

The capacitance, for instance, of a parallel plate capacitor, can be written as follows:

$$C = \epsilon_0 \epsilon_R A / d \tag{3.1}$$

Where,

ϵ_0 is the permittivity of space ($8.854 \times 10^{-12}\ F/m$)
ϵ_R is the dielectric constant of the material between the parallel plates
A is the area of the plates
d is the separation distance between the plates.

For a capacitor to work as a sensor, the capacitance should be the function of the quantity to be sensed (measurand). Therefore, the capacitance should change with the change in the measurand. This is possible when the measurand changes one of the parameters that define the capacitance. As per Eq. 3.1, if either of the distance between plates, area of the plates, or the dielectric constant of the material is a function of the measurand, the capacitance will change with respect to the quantity being sensed. In this way, a capacitor can be used as a sensor. For example, a relative humidity sensor can be designed by using a dielectric material, which is a function of the surrounding moisture (e.g., aluminum oxide or titanium dioxide) [7]. The change in the relative humidity will change the capacitance. The amount of relative humidity level can be determined by detecting the change in the capacitance using an electronic detection circuit. It should be noted that the Eq. 3.1 is valid only for parallel plate architecture [8]. The change in geometry will lead to different expressions for the capacitance with respect to the parameters; however, the principle of capacitive sensing remains the same.

Capacitive sensors provide numerous advantages such as no static power consumption, high resolution and sensitivity, and easy fabrication process. Another interesting aspect of capacitive sensors is that they can be used to detect metallic and non-metallic targets. On the other hand, capacitive sensors' capacitance is highly sensitive to changes in environmental conditions such as humidity and temperature. This may affect the performance of the sensor system [6].

3.1.1.2 Resistive Sensors

In the simplest term, the ability of a material to allow the flow of current is called resistivity (ρ) of the material with resistance (R). The relationship between the resistance and the resistivity of a conductor can be written as follows:

$$R = \rho \frac{l}{A} \tag{3.2}$$

where l is the length, and a is the cross-section area of the conductor. The property of a conductive material to control the current flow can be used to fabricate a sensor.

Similar to the capacitive sensing principle, in order to use a conductor as a resistive sensor, one of the parameters of Eq. 3.2 should change with respect to the measurand. Therefore, by modulating either the geometry factor or the specific resistivity, a resistor can be used as a sensor. For example, one of the most commonly used resistive sensors is the strain gauge. The basic strain gauge consists of a metallic foil pattern on an insulated flexible substrate. The applied force alters the geometry of the metallic foil, which changes the sensor's overall resistance. By detecting the change in the resistance, the value of the applied force can be detected [1].

3.1.1.3 Electrical Equivalent of R-C Sensors

The lossless sensor can be represented in an electrical equivalent form by either a capacitor in the case of the capacitive flexible sensor or a resistor in the case of the resistive sensors. However, the assumption is not valid for many sensing applications such as water level sensing, humidity sensing, and fingerprint sensing. The sensor's practical electrical equivalent model consists of a leakage shunt resistor and/or parasitic capacitor with the sensor element, as shown in Fig. 3.2. These non-idealities depend on many factors such as the improper structure on the substrate,

Fig. 3.2 Electrical equivalent
circuit for R-C sensors

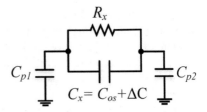

electrodes geometry, fabrication process, type of measurand, and influenced by environmental factors such as humidity and pollution. Furthermore, in some sensing applications, both capacitance and resistance are a function of the measurand and need to be simultaneously measured.

The equivalent electrical model of the impedance R-C flexible sensors is shown in Fig. 3.2. For leaky capacitive sensors, the capacitance C_x represents the sensor capacitor, and the resistor R_x is the shunt resistor due to the leakage current. Similarly, for resistive sensors, R_x represents the sensor resistor, and C_x is the shunt parasitic capacitor. In the case of the impedance sensors both R_x and C_x carry the information about the measurand. The capacitor C_{p1} and C_{p2} are the parasitic capacitors due to the connection of the electrodes and applicable for all sensors.

3.2   Signal Conditioning Circuits

The signal conditioning circuit modulates the sensors' output signal in such a way that it meets the requirement of the next stage (display or further processing). Based on the type of sensor, the signal conditioning circuits have at least one of the following operations: amplification, filtering, linearization, calibration, offset compensation, parasitic capacitance compensation, and leakage resistance compensation [9, 10].

3.3   Mismatch and Device Noise Reduction Techniques

The dominant error sources in low-frequency sensing applications are offset, offset drift, and $1/f$ noise. Offset is caused due to mismatches between the devices. Offset drift is caused due to temperature variation, and $1/f$ noise is caused due to the defects in the devices. Auto-zeroing and chopping are two commonly used techniques to address the issues of offset, offset drift, and $1/f$ noise.

3.3.1   Auto-Zeroing

Auto-zeroing is a two-phase discrete-time technique. The offset is sampled and stored in the auto-zero phase, and then the stored offset is subtracted from the input signal in the active phase. There are three basic topologies for auto-zeroing [11]: open-loop offset cancellation, closed-loop offset cancellation, and closed-loop offset cancellation using an auxiliary amplifier.

3.3.1.1   Open-Loop Offset Cancellation

An auto-zeroed amplifier with open-loop offset cancellation is shown in Fig. 3.3a. When f_{clk} is high, the amplifier is in auto-zeroing phase, where the amplified offset voltage is stored on capacitor C. When f_{clk} is low, the amplifier is in active phase, where the input V_{in} along with the offset voltage is amplified resulting in the cancellation of the offset voltage in the output V_{out}. This cancellation is possible

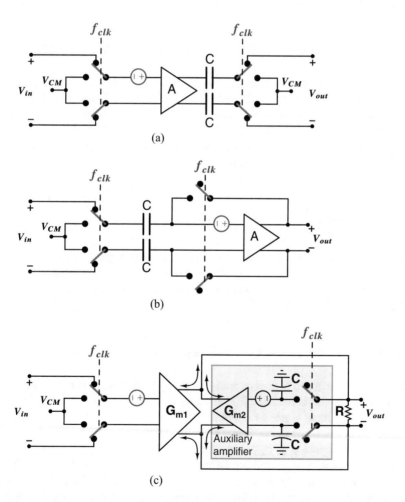

Fig. 3.3 (**a**) Open-loop offset cancellation (**b**) closed-loop offset cancellation (**c**) closed-loop offset cancellation using an auxiliary amplifier

because, in the auto-zero phase, the voltage on the capacitor is inverted to the case in the active phase. However, the open-loop offset cancellation technique limits the gain of the amplifier.

3.3.1.2 Closed-Loop Offset Cancellation

An auto-zeroed amplifier with closed-loop offset cancellation is shown in Fig. 3.3b. When f_{clk} is high, the amplifier is in auto-zeroing phase, where the offset voltage is stored on the capacitor C due to unity gain feedback configuration of the amplifier. When f_{clk} is low, the amplifier is in active phase, where the input V_{in} along with the stored offset voltage stored on the capacitor C results in the cancellation of the offset voltage at the input of the amplifier.

3.3.1.3 Closed-Loop Offset Cancellation Using an Auxiliary Amplifier

An auto-zeroed amplifier with closed-loop offset cancellation using an auxiliary amplifier is shown in Fig. 3.3c. In this configuration, the offset storage capacitor C is isolated from the signal path. When f_{clk} is high, the amplifier is in auto-zeroing phase, where the offset voltage of both the amplifiers is stored on the capacitor C. When f_{clk} is low, the amplifier is in active phase, where the auxiliary amplifier helps in the cancellation of the offset voltage of the amplifiers.

3.3.2 Chopping

The DC/low-frequency signal measurement accuracy is highly dependent on offset, offset drift, and $1/f$ noise. Chopping is a modulation technique that helps in the separation of the DC/low-frequency signal with the circuit non-idealities (offset, offset drift, and $1/f$ noise).

3.3.2.1 Voltage-Mode and Current-Mode Chopper Modulation Technique

The schematic diagram of the conventional chopping technique in the INA is shown in Fig. 3.4a. In the conventional technique, the noise separation is achieved by modulating the input signal to a higher frequency f_{mod} before the flicker noise is introduced (using Chp_1 in Fig. 3.4a). When the flicker noise is introduced due to the INA, the input signal is at f_{mod}, whereas the flicker noise is still at low frequency (near DC). Now to recover back the DC signal, the modulated signal is demodulated with the same frequency f_{mod} using a chopper (using Chp_2 in Fig. 3.4a) at the output of the INA. Moreover, Chp_2 also modulates the low-frequency flicker noise to a high-frequency f_{mod}. Thereby resulting in separation of flicker noise and the

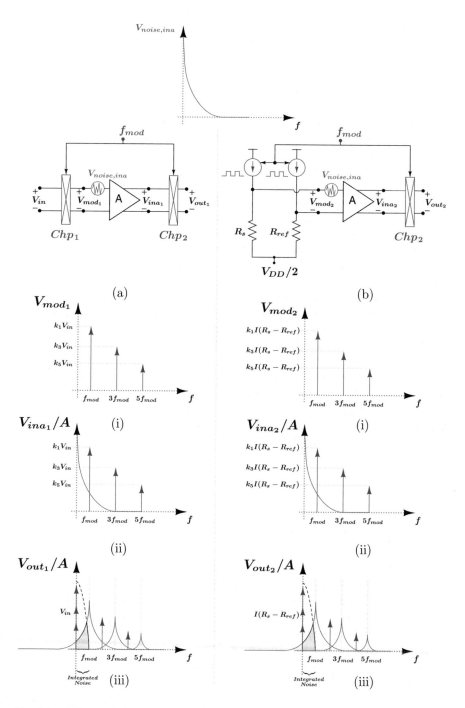

Fig. 3.4 (**a**) Chopped INA with modulation in the voltage domain [11]. (**b**) Chopped INA with modulation in the current domain [6]. (**a**) (i) Frequency response of V_{mod_1} due to V_{in} and Chp_1. (**b**) (i) Frequency response of V_{mod_2} due to square-wave current generator, R_s and R_{ref}. (**a**) (ii) Frequency response of V_{ina_1}. (**b**) (ii) Frequency response of INA V_{ina_2}. (**a**) (iii) Frequency response of V_{out_1}. (**b**) (iii) Frequency response of V_{out_2}

DC signal at the output of the INA. The higher the separation (higher f_{mod}), the better noise reduction is achieved at DC.

The chopping technique in Fig. 3.4b is based on the current modulation and reduces the flicker noise similar to the conventional technique. A detailed comparison between the utilized current-domain modulation and the conventional voltage-domain modulation scheme is shown in Fig. 3.4. In Fig. 3.4b, a square-wave current at frequency f_{mod} is used to excite the sensor. The square-wave current generators do the modulation in current domain. The modulated current then flows through the resistor and develop a modulated voltage with frequency f_{mod} across the input of the INA. This voltage is similar to having a modulated voltage signal (voltage domain), as shown in Fig. 3.4a.

Figure 3.4 compares the frequency spectrum at each node between Fig. 3.4a,b. Figure 3.4a represents the INA with DC input V_{in} and choppers Chp_1 and Chp_2 at input and output of the INA. The input chopper Chp_1 of the Fig. 3.4a modulated the DC input voltage signal V_{in} at the modulation frequency f_{mod} as shown in Fig. 3.4a,i. The chopper Chp_2 at the output of Fig. 3.4a demodulates the signal with the same frequency f_{mod}. This brings back the information at DC and the flicker noise get modulated to the higher frequency f_{mod} as shown in Fig. 3.4a(iii). Similarly, in the INA shown in Fig. 3.4b, the square-wave current generators do the modulation in current domain at the frequency f_{mod}. The modulated current signal is then converted into voltage using the sensor resistance. The output of the INA contains the sensor information and noise at the frequency f_{mod}. The chopper at the output of the INA demodulates the signal with the same frequency f_{mod} and recovers the sensor information as shown in Fig. 3.4a(iii). Therefore, the INA in Fig. 3.4b does suppress the flicker noise by the combined operation of the modulation in the current generator module, and the chopper at the INA output suppresses the flicker noise. The same is illustrated in Fig. 3.4.

3.3.2.2   Dual Chopping Technique

The Chopping technique helps separate the DC/low-frequency signal with the circuit non-idealities (offset, offset drift, and 1/f noise). The two conventional architectures are shown in Fig. 3.5a,b [11–13]. The reduction of $1/f$ noise in the conventional architectures is highly dependent on the chopping frequency f_{mod}. However, the higher the f_{mod}, the lower is the effect of $1/f$ noise on the measurement. Moreover, the higher the f_{mod}, the switch non-idealities dominate the error in the measurement. To address the problems in the conventional architecture (Fig. 3.5a,b), dual chopping technique/nested chopping technique can be used as shown in Fig. 3.5c,d [14, 15]. The chopper driven by f_{chop} helps in the reduction of the $1/f$ noise of the amplifier, whereas the chopper driven by f_{mod} is used for modulation and demodulation.

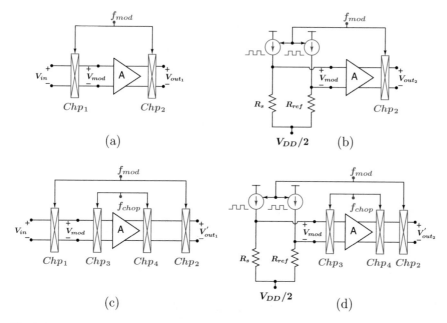

Fig. 3.5 (**a**) Chopped INA with modulation in the voltage domain. (**b**) Chopped INA with modulation in the current domain. (**c**) Chopped INA with modulation in voltage domain with dual chopping frequency operation [14]. (**d**) Chopped INA with modulation in the current domain with dual chopping frequency operation [15]

3.4 Interface Circuit Examples

3.4.1 *Interfacing for Resistive Sensors*

The signal conditioning circuits for the resistive sensors are mostly based on the Wheatstone bridge. The Wheatstone bridge provides a linear output with high sensitivity for strain gauge resistive sensors in full-bridge configuration. However, for single element resistive sensors, the Wheatstone bridge's output is non-linear with respect to the change in the sensor resistance. This affects the sensitivity and dynamic range of the measurement. Therefore, the Wheatstone bridge based signal conditioning circuit is not preferred for single element resistive sensors [1].

3.4.1.1 Auto-Nulling Technique for Resistive Sensors

Auto-nulling based techniques are utilized to linearize the output of the Wheatstone bridge circuits. An integrator based auto-nulling loop for the single element Wheatstone bridge is presented in [16, 17]. The circuit utilizes an analog multiplier based voltage-controlled resistor (VCR) in one arm of the Wheatstone bridge. The

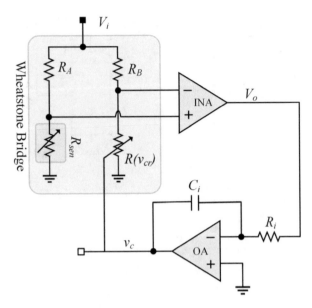

Fig. 3.6 Schematic diagram of the auto-nulling based signal conditioning circuit for bridge based resistive sensors

auto-nulling loop controls the resistance of the VCR and keeps the bridge in a balanced condition. The incremental change in the sensor resistance is proportional to the variation in the auto-nulling loop's output voltage. This variation is linear with respect to ΔR.

The simplified schematic of the Wheatstone bridge with integrator based auto-nulling is shown in Fig. 3.6. The sensor is represented by R_{sen} and the VCR is represented by R_{vcr}. The output of each arm of the Wheatstone bridge is amplified by the instrumentation amplifier and converted into a single-ended voltage. The voltage V_o is applied to the input of the integrator. The integrator works as an integral controller and controls the VCR. The integrator output settles at the balanced condition when V_o becomes zero. The value of the sensor resistance $R_{s}en$ at balanced condition can be written as follows:

$$R_{sen} = \frac{R_A}{R_B} R_{vcr} \tag{3.3}$$

The Eq. 3.3 shows that the sensor resistance is linearly related to the R_{vcr}. By implementing the linear VCR, such as the one implemented in [17] using an analog multiplier and resistor, a linear R_{sen} with respect to the voltage V_c can be obtained.

In addition, the auto-nulling loop enhances the robustness of the signal conditioning circuits against the effect of variation of the circuit parameters on the measurement of the sensor resistance. The output voltage of the auto-nulling circuit reported in [17] only depends on the bridge components and the output voltage at

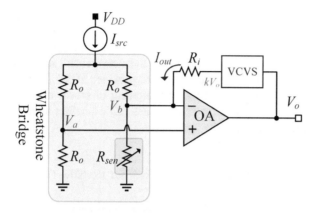

Fig. 3.7 Schematic diagram of the bridge auto-nulling technique using the negative feedback loop

the balanced condition. Therefore, any variations in any other circuit parameters and the supply voltage and frequency will not affect the output voltage.

Another auto-nulling technique for Wheatstone bridge based single element resistive sensors reported in [18] is shown in Fig. 3.7. The circuit is based on a negative feedback loop, which keeps the voltage V_a and V_b at a fixed potential. Any variations in the ΔR is compensated by the negative feedback loop by providing additional current I_{out}. This additional current is supplied by the voltage-controlled-voltage-source (VCVS) block placed in the feedback of the amplifier. The current Iout is proportional to ΔR. The value of ΔR can be calculated by measuring the voltage V_o. The value of ΔR can be written as follows:

$$R_{sen} = \frac{R_o}{R_i}\left(\frac{2kV_o}{I_{scr}} - R_o\right) \tag{3.4}$$

where k is the gain-factor of the VCVS.

Therefore, as per 1.4, the incremental change of ΔR in the sensor resistance can be obtained by measuring the voltage V_o. The auto-nulling circuits help in linearizing the output of the Wheatstone bridge circuits for single element resistive sensors. However, the effect of the thermoelectric offset (if excitation signal is DC) and parasitic capacitances (if the excitation signal is AC) needs to be addressed.

The conventional resistance measurement system is mostly based on DC excitation. The auto-nulling approach shown in Fig. 3.6 is also based on DC excitation. However, the system based on DC excitation suffers from the thermoelectric effect due to the interconnection between the sensor and the signal conditioning. The INA further amplifies this thermoelectric offset and dominantly affects the system's accuracy and measurement range. Therefore, AC excitation is preferred to avoid the thermoelectric offset voltage. The sinusoidal voltage source is mostly preferred for the excitation signal due to its single tone frequency and no additional harmonics.

The AC excitation signal removes the effect of the DC offset voltage generated due to the thermoelectric effect by filtering the DC components. However, AC excitation causes the parasitic capacitance in parallel with the sensor resistance to contribute to the output voltage. This affects the measurement accuracy of the sensor resistance. Therefore, the provision should be provided to compensate for the effect of such parasitic capacitance on the measurement of sensor resistance. The auto-nulling based circuit reported in [18] utilizes AC excitation signal; however, the output suffers from the parasitic capacitance.

3.4.1.2   Relaxation Oscillator Based Circuit for Resistive Sensors

Relaxation Oscillator based signal conditioning circuits are widely used for resistive sensors. Thanks to the inherent properties such as the quasi-digital output and high noise-immunity, the relaxation oscillator based circuits are preferred for signal conditioning circuit [19].

The circuit diagram of the basic relaxation oscillator circuit is shown in Fig. 3.8. The circuit consists of an integrator and a Schmitt trigger. The output of the Schmitt trigger is applied to the input of the integrator. The current flows through the resistor R_x charges and discharges the capacitor C_i based on the polarity of the voltage $V_x(t)$. The Schmitt trigger sets the threshold voltage at the output of the integrator. Once the voltage $V_R(t)$ reaches the threshold voltage, the output of the Schmitt trigger $V_X(t)$ switch reverses the polarity of the current in the capacitor C_i. This process is repetitive, and we obtain a square-wave signal at the output whose frequency is proportional to the value of sensor resistance R_x. The expression for the frequency of the basic relaxation oscillator circuit can be written as follows [20]:

$$f = 4\frac{R_2}{R_1}R_x C_i \tag{3.5}$$

The circuit shown in Fig. 3.8 is suitable for single element floating resistive sensors. The relaxation oscillator based circuit is modified to extend the application

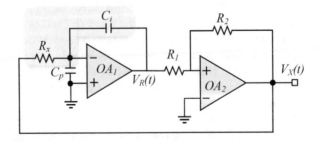

Fig. 3.8 Basic relaxation oscillator circuit

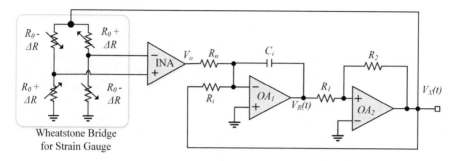

Fig. 3.9 Relaxation oscillator based resistance-to-time converter circuit for strain gauge resistive sensors

for different sensing configurations such as the full-bridge sensor, grounded resistive sensors, and resistive sensors with parasitic capacitance [19, 21–24].

An extended interface circuit for strain gauge using relaxation oscillator is shown in Fig. 3.9. The output frequency of the circuit is proportional to the change in the sensor resistor ΔR, and can be derived as follows:

$$f = 4\frac{R_2}{R_1} R_x C_i \tag{3.6}$$

3.4.2 Interfacing for Capacitive Sensors

Capacitive sensors have various unique features, such as high sensitivity, small size, and no static power consumption. Because of these features, the capacitive sensors are widely used in portable battery powered systems for sensing parameters such as displacement, solid contaminants in gas pipelines, humidity, moisture, pressure, and even in wireless implant systems for bio-medical applications [25–27].

The design of the capacitive sensors can be classified into two categories: grounded and floating. In grounded capacitive sensors, one terminal of the sensor is always connected to the ground, whereas in floating capacitive sensors, both electrodes can be connected to any terminal of the signal conditioning circuit as per the requirement of sensor [28].

3.4.2.1 Phase-Sensitive-Detection Technique for Capacitance Measurement

The phase-sensitive-detection (PSD) technique for the capacitive sensors is based on demodulating the sensor component using a reference quadrature-phase shifted signal. Schematic diagram of the basic PSD based interface circuit is shown in Fig. 3.10. The schematic consists of a leaky capacitive sensor. The circuit is suitable

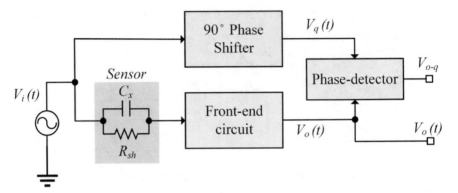

Fig. 3.10 Schematic diagram of the phase-sensitive-detection based capacitance measurement circuit

for both the leaky and lossless capacitive sensors. The front-end circuit converts the sensor parameter and leakage resistance into equivalent voltage signal $V_o(t)$. The voltage $V_o(t)$ is demodulated using a quadrature-phase shifted reference signal $V_q(t)$. The demodulation is performed either using a single-pole-double through switch or using a multiplier. The demodulation mitigates the leakage resistance R_x to high frequency and brings the sensor capacitance C_x components to DC. The low-pass-filter (LPF) filters out the high-frequency components, and the output voltage of LPF is proportional to the sensor capacitance.

Considering $V_i(t) = V_m \sin \omega t$ where V_m is the amplitude and ω is the angular frequency of the input excitation signal. The quadrature phase shifted signal $V_q(t)$ is shifted by a phase $\phi = 90°$. Suppose the output of the analog front end is given as follows:

$$V_o(t) = V_m[\alpha G_{sh} \sin \omega t + \beta C_x \cos \omega t)] \tag{3.7}$$

When an analog multiplier is used as the phase-detector, the voltage $V_o(t)$ and $V_q(t)$ multiples and provides the output voltage $V_{(o-q)}$. The voltage $V_{(o-q)}$ can be written as follows:

$$V_{o-q} = \frac{V_o(t) \times V_q(t)}{mul} = \frac{(V_m[\alpha G_{sh} \sin \omega t + \beta C_x \cos \omega t) \times V_m sin(\omega t + \pi/2)]}{V_m ul} \tag{3.8}$$

where V_{mul} is the attenuation factor of analog multiplier. The voltage $V_{(o-q)}$ is passed through a low-pass filter, which filtered out the high-frequency components. The output of the low-pass filter can be written as follows. The output voltage of such signal conditioning circuits is sensitive to the amplitude and frequency of the input excitation signal. Any variation and/or fluctuation in the amplitude and frequency of the input signal affects the accuracy of the measurement of sensor capacitance.

3.4.2.2 Auto-Nulling Based Signal Conditioning Circuit for Capacitive Sensors

The auto-nulling based signal conditioning circuits are used to convert the sensor capacitance into equivalent output voltage for lossless and leaky sensors. The auto-nulling based signal conditioning circuit utilizes a negative feedback loop that provides robustness in the measurement of sensor component against the variation in the other sources of non-idealities such as the variation in the passive component values and excitation signal. This improves the accuracy of the measurement [29].

The auto-nulling bridge based capacitance-to-voltage converter for capacitive sensor, reported in [23], is shown in Fig. 3.11. The circuit consists of a De Sauty bridge followed by an INA and a phase-sensitive-detector, and an auto-nulling loop. The quadrature component of the bridge is separated by the analog multiplier based PSD. The De Sauty bridge utilizes a voltage-controlled-resistor. The auto-nulling loop forces the multipliers' output V_{mix} to become zero by controlling the resistance R_{vcr} of the bridge. At null condition, the output V_o is proportional to the sensor capacitance C_{sen}. The circuit shown in Fig. 3.11 is sensitive to the leakage resistance and parasitic capacitance of the sensor. To compensate the effect of such non-idealities, an improved auto-balancing circuit is reported for sensor capacitance measurement [30]. The conventional De Sauty bridge circuit is modified

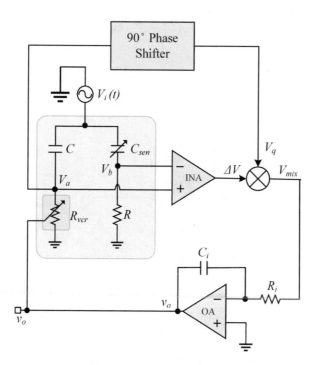

Fig. 3.11 Auto-nulling based capacitance-to-voltage converter circuit

by introducing an inverting amplifier as shown in Fig. 3.11. The circuit consists of two-branches: a sensor branch and a reference branch. The main idea of the circuit is based on generating an accurate quadrature phase-shifted signal and demodulating the sensor current with the quadrature phase-shifted signal. The demodulated sensor current is then compensated using a control loop, which generates a reference current signal equal to the sensor current using a feedback loop and a voltage-controlled capacitor.

The modified bridge circuit, shown in Fig. 3.12, is formed by an inverting amplifier A_1, the leaky sensor, and a voltage-controlled capacitor. A capacitor C_s and an analog multiplier M_3 are used for realizing the voltage-controlled capacitor. The current I_X flows through the leaky sensor and I_s flows through the reference capacitor C_s. The effective current I_c is converted into voltage using a current-to-voltage converter. The voltage $V_a(t)$ is demodulated using a reference quadrature-phase shifted signal. The output of the multiplier contains the information of the sensor capacitance at DC and leakage resistance on the high frequency. Low-pass filter was used to mitigate the effect of leakage resistance. The output of the LPF is proportional to the sensor capacitance. The integrator integrates the LPF output till the voltage V_L becomes zero. The proportional voltage V_{sens} generates the current I_s. The output V_{sens} can be derived as follows [30, 31]:

$$V_{sens} = C_x / C_s V_{m3} \tag{3.9}$$

where V_{m3} is the attenuation factor of analog multiplier M_3.

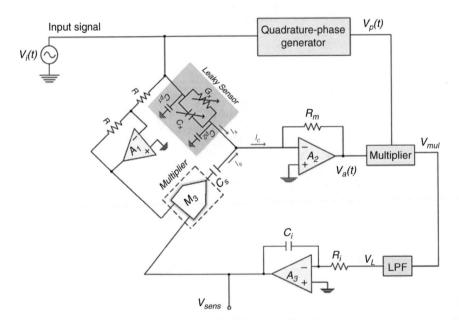

Fig. 3.12 Improved auto-balancing bridge based signal conditioning circuit for leaky capacitive sensors

3.4.3 Interfacing for Impedance R-C Sensors

The lossless sensor can be represented in an electrical equivalent form by either a capacitor in the case of the capacitive sensor or a resistor in the case of the resistive sensors. However, the assumption is not valid for many sensing applications. The sensor's practical electrical equivalent model consists of a leakage shunt resistor and/or parasitic capacitor with the sensor element. In such sensors, both capacitance and resistance of the sensor need to be measured continuously.

3.4.3.1 Phase-Sensitive-Detection Based Impedance-to-Voltage Converter

The phase-sensitive-detection based circuits are widely used for the complex impedance R-C sensors. The schematic diagram of a PSD based impedance-to-voltage converter circuit for impedance R-C sensors is shown in Fig. 3.13. The circuit is based on separating the in-phase and quadrature components of the R-C sensor using phase-detectors and reference signals. The capacitance of the impedance R-C sensors is separated by modulating the output $V_O(t)$ of the front-end circuit with a reference quadrature-phase shifted signal $V_q(t)$. The modulation is performed using a phase-detector. The modulation operation shifts the resistive components to higher frequency, which can be filtered out using a low-pass filter. Similarly, the resistance of the impedance R-C sensor is measured by modulating the output $V_o(t)$ using a reference in-phase signal $V_i(t)$. The modulated output is filtered using a low-pass filter to remove the effect of quadrature components [32].

A PSD based impedance-to-voltage converter circuit for impedance R-C sensors is reported in [33]. The circuit separates the in-phase and quadrature components of the sensor with high accuracy. However, the output of the circuit reported in [27] is sensitive to the amplitude and frequency of the excitation signal. Furthermore,

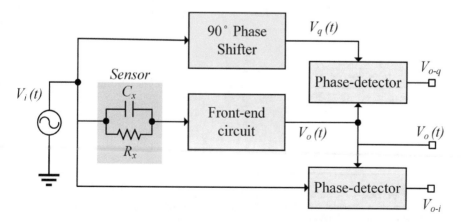

Fig. 3.13 Phase-sensitive-detector based impedance-to-voltage converter

the circuit is also sensitive to any variations in passive component values and non-idealities of the active components. Auto-nulling based impedance-to-voltage converter for impedance R-C sensors is reported in [34]. The auto-nulling circuit utilizes a negative feedback loop that compensates the effect of the other circuit components and variation on the measurement of sensor parameters. This enhances the accuracy and robustness of the signal conditioning circuit.

3.4.4 Interface Excitation Techniques

The measurement accuracy depends on the type of modulation signal and the type of synchronous demodulation scheme. Four different combinations of modulation and demodulation scheme are shown in Fig. 3.14. The sine-wave excitation with multiplier based demodulation response is shown in Fig. 3.14a. In principle, this approach can provide the best linearity. However, the performance degrades in

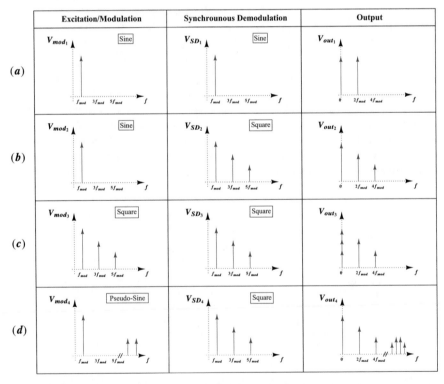

Fig. 3.14 (**a**) Sine-wave excitation with multiplier based demodulation. (**b**) Sine-wave excitation with square-wave demodulation. (**c**) Square-wave excitation with square-wave demodulation. (**d**) Pseudo-sine wave excitation with square-wave demodulation

the implementation of multiplier due to the nonlinearity, mismatch, and offset of the multiplier. The problems associated with the multiplier can be avoided by using sine-wave excitation with square-wave demodulation scheme. The frequency response of the sine-wave excitation with square-wave demodulation is shown in Fig. 3.14b. The implementation of the square-wave demodulator is simple, and it avoids the issues associated with the multiplier. The sine-wave generation with high linearity is power inefficient. One approach can be to use square-wave excitation with square-wave demodulation. The frequency response of the square-wave excitation with square-wave demodulation is shown in Fig. 3.14c. The square-wave excitation is power efficient; however, the output signal linearity gets severely affected. Moreover, the linearity improves in a specific case using DRO (differential and ratiometric operation) technique [15]. To improve the linearity with low-power consumption a pseudo-sine waveform synthesizer using the adaptive quantization DAC control can be used. This helps in improving the linearity without increasing the power consumption. The frequency response of the pseudo sine-wave excitation with square-wave demodulation is shown in Fig. 3.14d [12].

3.5 Low-Power Low-Noise CMOS Instrumentation Amplifiers

The instrumentation amplifier (INA) is a critical block in any measurement system with AC excitation. High CMRR (to reject supply interferences and square-wave common-mode input signals) and low input-referred noise are extremely important criteria for the design of INA. Commonly used INA architectures in the literature are discussed in the subsequent subsections.

3.5.1 3-Op-Amp Based Instrumentation Amplifier

The 3-op-amp based INA as shown in Fig. 3.15 is widely used in the Commercial off-the-shelf (COTS) ICs for its high input impedance and high CMRR. The gain of the INA is expressed as R_f/R_i. However, the high CMRR is achieved due to expensive resistor trimming process. Moreover, the noise-power trade-off is not good. The dominant noise contributors are the input amplifiers and the feedback resistor R_f. Hence, for low noise, the INA needs large power input stages to lower the input differential pair noise and large power output stages to drive R_f [12].

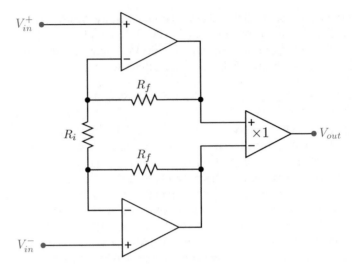

Fig. 3.15 3-Op-amp based instrumentation amplifier

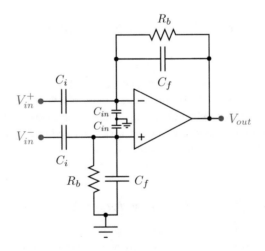

Fig. 3.16 Capacitive feedback instrumentation amplifier

3.5.2 Capacitive Feedback Instrumentation Amplifier

Capacitive feedback based INA as shown in Fig. 3.16 is also a common INA topology for a smaller area, but it suffers from low CMRR due to the requirement of matched capacitor C_f and C_i. The noise of INA is high due to input parasitic capacitor C_{in}, which can be expressed as Eq. 3.10. The input impedance of the INA is low due to capacitor C_i [35]. The gain of the INA can be expressed as C_i/C_f. For area efficiency, large bias resistances R_b are realized using pseudo resistors.

However, pseudo resistors are sensitive to PVT variations. Moreover, a large bias resistor R_b, results in slowing down the settling time of the INA [12].

$$v_{ni}^2 = \left(\frac{C_i + C_f + C_in}{C_i}\right)^2 \cdot v_{nia}^2 \qquad (3.10)$$

3.5.3 Current-Mode Instrumentation Amplifiers

Current-Mode INA is a common INA topology for high CMRR, which does not depend on resistor matching and maintaining low-power consumption, as shown in Fig. 3.17. The noise is determined by the input stage and the resistor R_i. Compared to the 3-op-amp based INA, the input stage drives resistor R_i. Hence, CMIA does not require high power output stages [12]. The input voltage to the INA is buffered and then converted to equivalent current using the resistor R_i. This equivalent current is mirrored and fed to a large resistor R_o, generating output with a gain of R_o/R_i.

The G_m stage, mirror stage, and TI stage can be implemented as shown in Fig. (Fig. 3.18). The G_m stage includes the transistors $M_1 - M_7$ and $M_1' - M_7'$, which constitutes the flipped voltage follower (FVF) buffer. These FVF buffers convert the input signals $V_{in}^+ - V_{in}^-$ into an equivalent current with the resistor R_i. The current flowing through the resistor R_i and the transistors M_2 and M_2' (G_m stage) is mirrored using transistors M_2, M_2', M_8, and M_8' (TI stage). Hence, the mirrored current flow through the resistor R_o, generating an amplified voltage R_o/R_i.

Fig. 3.17 Current-mode instrumentation amplifier

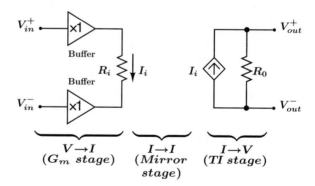

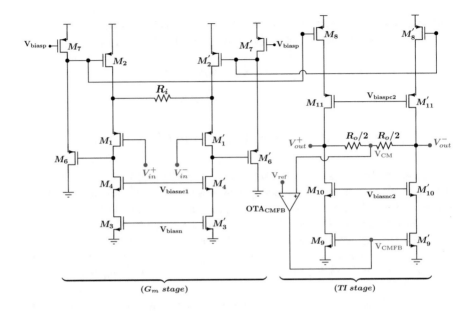

Fig. 3.18 Current-mode instrumentation amplifier (transistor level implementation)

References

1. L. Baxter, *Capacitive Sensors Design and Application* (Wiley-IEEE Press, New Delhi, 1997)
2. S. Malik, M. Ahmad, M. Punjiya, A. Sadeqi, M.S. Baghini, S. Sonkusale, Respiration monitoring using a flexible paper-based capacitive sensor, in *2018 IEEE SENSORS* (2018), pp. 1–4
3. S. Malik, L. Somappa, M. Ahmad, S. Sonkusale, M.S. Baghini, A fringing field based screen-printed flexible capacitive moisture and water level sensor, in *2020 IEEE International Conference on Flexible and Printable Sensors and Systems (FLEPS)* (2020), pp. 1–4
4. M. Ahmad, S. Malik, L. Somappa, S. Sonkusale, M.S. Baghini, A flexible dry ECG patch for heart rate variability monitoring, in *2020 IEEE International Conference on Flexible and Printable Sensors and Systems (FLEPS)* (2020), pp. 1–4
5. L. Somappa, S. Malik, M. Ahmad, K.M. Ehshan, A.M. Shaikh, K.M. Anas, S. Sonkusale, M.S. Baghini, A 3d printed robotic finger with embedded tactile pressure and strain sensor, in *2020 IEEE International Conference on Flexible and Printable Sensors and Systems (FLEPS)* (2020), pp. 1–4
6. J. Fraden, *Handbook of Modern Sensors: Physics, Designs, and Applications* (Springer, New Delhi, 2010)
7. X. Zhao, X. Chen, X. Yu, P. Du, N. Li, X. Chen, Humidity-sensitive properties of tio2 nanorods grown between electrodes on au interdigital electrode substrate. IEEE Sensors J. **17**(19), 6148–6152 (2017)
8. F. Reverter, X. Li, G.C. Meijer, Liquid-level measurement system based on a remote grounded capacitive sensor. Sens. Actuators, A **138**(1), 1–8 (2007)
9. A.D. Marcellis, G. Ferri, *Analog Circuit and System for Voltage-Mode and Current-Mode Sensor Interfacing Application* (Springer, New Delhi, 2011)

10. L. Somappa, S. Malik, S. Aeron, S. Sonkusale, M.S. Baghini, High resolution frequency measurement techniques for relaxation oscillator based capacitive sensors. IEEE Sensors J. **21**(12), 13 394–13 404 (2021)
11. C. Enz, G. Temes, Circuit techniques for reducing the effects of op-amp imperfections: autozeroing, correlated double sampling, and chopper stabilization. Proc IEEE **84**(11), 1584–1614 (1996)
12. N. Van Helleputte, M. Konijnenburg, J. Pettine, D.-W. Jee, H. Kim, A. Morgado, R. Van Wegberg, T. Torfs, R. Mohan, A. Breeschoten, H. de Groot, C. Van Hoof, R.F. Yazicioglu, A 345 μw multi-sensor biomedical SoC with bio-impedance, 3-channel ECG, motion artifact reduction, and integrated DSP. IEEE J. Solid State Circuits **50**(1), 230–244 (2015)
13. N. Van Helleputte, S. Kim, H. Kim, J.P. Kim, C. Van Hoof, R.F. Yazicioglu, A 160 μa biopotential acquisition IC with fully integrated IA and motion artifact suppression. IEEE Trans. Biomed. Circuits Syst. **6**(6), 552–561 (2012)
14. A. Bakker, K. Thiele, J. Huijsing, A CMOS nested-chopper instrumentation amplifier with 100-nv offset. IEEE J. Solid State Circuits **35**(12), 1877–1883 (2000)
15. M. Ahmad, S. Malik, S. Dewan, A.K. Bose, D. Maddipatla, B.B. Narakathu, M.Z. Atashbar, M.S. Baghini, An auto-calibrated resistive measurement system with low noise instrumentation ASIC. IEEE J. Solid State Circuits **55**(11), 3036–3050 (2020)
16. G. Ferri, V. Stornelli, A. De Marcellis, A. Flammini, A. Depari, Novel CMOS fully integrable interface for wide-range resistive sensor arrays with parasitic capacitance estimation. Sensors Actuators B Chem. **130**(1), 207–215 (2008). Proceedings of the Eleventh International Meeting on Chemical Sensors IMCS-11
17. A. De Marcellis, G. Ferri, P. Mantenuto, A novel 6-decades fully-analog uncalibrated Wheatstone bridge-based resistive sensor interface. Sensors Actuators B Chem. **189**, 130–140 (2013). Selected Papers from the 26th European Conference on Solid-State Transducers
18. E. Alnasser, A novel fully analog null instrument for resistive Wheatstone bridge with a single resistive sensor. IEEE Sensors J. **18**(2), 635–640 (2018)
19. V. Ferrari, C. Ghidini, D. Marioli, A. Taroni, Oscillator-based signal conditioning with improved linearity for resistive sensors. IEEE Trans. Instrum. Meas. **47**(1), 293–298 (1998)
20. T. Islam, L. Kumar, Z. Uddin, A. Ganguly, Relaxation oscillator-based active bridge circuit for linearly converting resistance to frequency of resistive sensor. IEEE Sensors J. **13**(5), 1507–1513 (2013)
21. A. Depari, A. Flammini, E. Sisinni, A. De Marcellis, G. Ferri, P. Mantenuto, Fast, versatile, and low-cost interface circuit for electrochemical and resistive gas sensor. IEEE Sensors J. **14**(2), 315–323 (2014)
22. Z. Ignjatovic, M. Bocko, An interface circuit for measuring capacitance changes based upon capacitance-to-duty cycle (CDC) converter. IEEE Sensors J. **5**(3), 403–410 (2005)
23. K. Mochizuki, K. Watanabe, T. Masuda, M. Katsura, A relaxation-oscillator-based interface for high-accuracy ratiometric signal processing of differential-capacitance transducers. IEEE Trans. Instrum. Meas. **47**(1), 11–15 (1998)
24. S. Malik, M. Ahmad, L. S, T. Islam, M.S. Baghini, Impedance-to-time converter circuit for leaky capacitive sensors with small offset capacitance. IEEE Sensors Lett **3**(7), 1–4 (2019)
25. Y. Jung, Q. Duan, J. Roh, A 17.4-b delta-sigma capacitance-to-digital converter for one-terminal capacitive sensors. IEEE Trans. Circuits Syst. Express Briefs **64**(10), 1122–1126 (2017)
26. B. George, V.J. Kumar, Analysis of the switched-capacitor dual-slope capacitance-to-digital converter. IEEE Trans. Instrum. Meas. **59**(5), 997–1006 (2010)
27. S. Malik, Q. Castellví, L. Becerra-Fajardo, M. Tudela-Pi, A. García-Moreno, M.S. Baghini, A. Ivorra, Injectable sensors based on passive rectification of volume-conducted currents. IEEE Trans. Biomed. Circuits Syst. **14**(4), 867–878 (2020)
28. R. Nojdelov, S. Nihtianov, Capacitive-sensor interface with high accuracy and stability. IEEE Trans. Instrum. Meas. **58**(5), 1633–1639 (2009)

29. P. Mantenuto, A. De Marcellis, G. Ferri, Novel modified De-Sauty autobalancing bridge-based analog interfaces for wide-range capacitive sensor applications. IEEE Sensors J. **14**(5), 1664–1672 (2014)
30. S. Malik, L. Somappa, M. Ahmad, M.S. Baghini, An-c2v: an auto-nulling bridge-based signal conditioning circuit for leaky capacitive sensors. IEEE Sensors J. **20**(12), 6432–6440 (2020)
31. S. Malik, K. Kishore, T. Islam, Z.H. Zargar, S. Akbar, A time domain bridge-based impedance measurement technique for wide-range lossy capacitive sensors. Sens. Actuators, A **234**, 248–262 (2015)
32. S. Malik, L. Somappa, M. Ahmad, T. Islam, M.S. Baghini, An accurate digital converter for lossy capacitive sensors. Sens. Actuators, A **331**, 112958 (2021)
33. A.U. Khan, T. Islam, B. George, M. Rehman, An efficient interface circuit for lossy capacitive sensors. IEEE Trans. Instrum. Meas. **68**(3), 829–836 (2019)
34. S. Malik, M. Ahmad, L. Somappa, T. Islam, M.S. Baghini, An-z2v: autonulling-based multimode signal conditioning circuit for RC sensors. IEEE Trans. Instrum. Meas. **69**(11), 8763–8772 (2020)
35. R. Harrison, C. Charles, A low-power low-noise CMOS amplifier for neural recording applications. IEEE J. Solid State Circuits **38**(6), 958–965 (2003)

Chapter 4
Data Converters for Wearable Sensor Applications

4.1 Introduction

Data converter circuits and systems are an essential block that interfaces the analog world of sensors with the digital world of signal processing. Data converter circuits essentially convert incoming signal information in voltage, current, capacitance, resistance, and impedance to digital signals for further digital signal processing. Wearable applications involve capturing and processing of analog information in the form of various bio-potential signals as shown in Fig. 4.1, body temperature, or any other body parameters in the form of change in resistance, capacitance, or impedance. The analog information in the voltage and current form can be converted to a digital signal using an analog to digital converter (ADC) and the analog information in the capacitance, resistance, and impedance form can be directly digitized using direct-digital converter (DDC). Data conversion is a two-step process where the analog information is first discretized in time, termed as sampling, and then quantized in amplitude (voltage, current, capacitance, resistance, or impedance), termed quantization. In the last several years, there has been considerable research on digital converter circuits for a wide range of applications. For wearable biomedical applications, the need is to record and monitor various bio-potential signals like electrocardiogram (ECG), electrooculography (EOG), electroencephalogram (EEG), electromyography (EMG), and axon action potentials (AAP). These bio-potential signal bandwidths can range from near dc to about 10 kHz with varying voltage levels ranging from a few uV to 100 mV, as illustrated in Fig. 4.1. Owing to the very low-voltage levels of the signals, an analog signal conditioning circuit is essential at the front end to suppress the noise and amplify the signal in the biomedical band, as shown in Fig. 4.1. A variety of these signal

Dr. Maryam Shojaei Baghini (IIT Bombay), Dr. Laxmeesha Somappa (IIT Bombay) and Dr. Shahid Malik (IIT Delhi) contributed to this chapter.

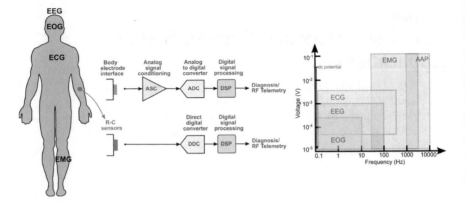

Fig. 4.1 Illustration of bio-signal acquisition system and the frequency spectrum of bio-potential signals

conditioning circuits were discussed in the previous chapter. This chapter focuses on the design of ADC for the bio-potential acquisition applications. Specifically, for portable and implantable medical systems where power consumption and energy efficiency of the system are of the highest importance. Specifically, a class of ADCs called delta-sigma modulators (DSM) will be discussed for very low-voltage and low-power multichannel bio-potential acquisition systems.

In addition, the flexible sensors are being used in a variety of different applications such as moisture, humidity, and pressure [1–3]. Moreover, in addition to the bio-potential signals, body sensor based systems are also getting wide attention for recording physiological information. Various sensors are employed to convert the physiological information into electrical signals. For instance, capacitive pressure sensors measure the blood pressure by converting the pressure change into the capacitance change. Similarly, resistive temperature sensors are employed to convert the body temperature into an electrical signal [4, 5].

In addition to physiological signal monitoring from the body, sensors have also been used to mimic the human skin for prosthetic applications. Skin is the largest organ of the human body and acts as a physical interface between the environment and the body. Researchers have developed flexible electronic skins (e-skin), which is basically a human skin-inspired electronic device applied to the body with human-like sensing capabilities. Often the output of such sensors is very weak and noisy—and therefore needs pre-processing or conditioning before it can be useful for the load (decision maker). Traditionally, appropriate signal conditioning circuits are employed to convert the output of such sensors into equivalent electrical signals. The signal conditioning circuits may have several stages, such as amplification, filtering, and linearization [6].

4.2   Low-Voltage ADC for Bio-Potential Acquisition

Flexible biomedical sensors are used for monitoring physiological parameters for healthcare applications [7, 8]. Demands for the simultaneous monitoring and recording of various bio-potential signals have been increased for a comprehensive study and implementation of numerous portable and implantable electronic medical systems. The most common bio-potential signal bandwidths can range from near dc to about 10 kHz with varying voltage levels ranging from a few mV to 100 mV [9]. The recent advances in microelectronics technology have facilitated fast growth in the number of parallel recording channels for simultaneous recording of the bio-potential signals [10]. From a circuit design perspective, implantable recording systems with parallel recording channels present challenges in the form of high density (compact circuits) and very low-power consumption. The systems generally exhibit very low-power consumption characteristics since they must operate for months or years without battery replacement. Reducing the power dissipation also helps reduce the risk of damaging surrounding tissues due to the dissipated heat [11]. Since the dynamic power is proportional to the square of the supply voltage VDD, reducing the operating VDD is one of the most effective ways to reduce the dynamic power consumption. Monitoring and recording biomedical signals of the human body require converting multichannel analog bio-potential signals into digital signals through ADCs. Compact, very low-power (operating at a low voltage), and energy-efficient ADCs are essential for the longevity of portable and implantable multichannel biomedical systems.

Conventional bio-potential acquisition and recording systems involving multiple channels consist of a dedicated low-noise amplifier (LNA) and a programmable gain amplifier (PGA) per channel, followed by an analog multiplexer for time-division multiplexing of the channels on to a single successive-approximation register (SAR) analog-to-digital converter (ADC) as illustrated in Fig. 4.2. This architecture provides an advantage in terms of a compact design owing to a single ADC required for M-channels resulting in an effectively reduced area per channel. However, with expanding channel counts, the architecture suffers from poor effective energy efficiency per channel, i.e., for large channel data acquisition and recording, the power consumption of the multiplexed architecture consumes more power than a fully parallel architecture with a dedicated ADC per channel [11–16], illustrated in Fig. 4.3. With large number of channels, the ADC

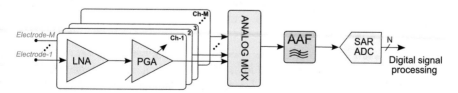

Fig. 4.2   A conventional multichannel bio-potential recording system with a time-multiplexed SAR ADC for M channels

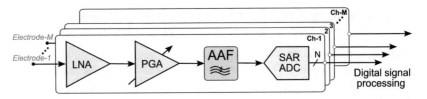

Fig. 4.3 A multichannel bio-potential acquisition system with a dedicated SAR ADC per channel

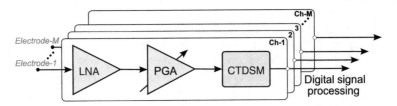

Fig. 4.4 A multichannel bio-potential acquisition and recording system with a continuous-time delta-sigma modulator

sampling time is effectively shortened due to the multiplexing leading to larger power consumption by the buffer/AAF driving the ADC. This increased power consumption results in a large overall system power consumption compared to the non-multiplexed architecture [14, 15]. Further, the routing parasitics due to the multiplexing significantly increase the load capacitance of the front-end blocks, which increases the overall system power dissipation [14].

The alternate conventional multichannel bio-potential recording system with a dedicated ADC per channel is illustrated in Fig. 4.3. A conventional SAR ADC with a moderate resolution of 8 or 9 bits is widely adopted since they offer good energy efficiency and moderate accuracy. However, the anti-aliasing filter (AAF)/buffer needs to drive a capacitor bank in the SAR ADC, leading to sub-optimal power and area usage [17]. Moreover, while the conventional SAR ADCs achieve good energy efficiency, the matching requirement of the capacitor array puts a fundamental limit on the area [17, 18]. An alternate approach is to replace the SAR ADC with a continuous-time $\Delta\Sigma$ modulator (CTDSM), which eliminates the need for an AAF due to the implicit anti-aliasing feature of a CTDSM [17] as illustrated in Fig. 4.4.

The focus of this section of the chapter will be the architectural choices for $\Delta\Sigma$ modulator (DSM) based bio-potential acquisition and recording systems. The architectures target a bandwidth of 10 kHz to cover a wide range of bio-potential signals. The DSMs discussed will have a very compact area typically less a 0.1 mm^2/channel. This stringent criterion on the area limits the use of large on-chip resistors and capacitors. Further, due to the limited battery power availability in portable bio-medical devices, and to avoid tissue damage in implantable applications, the DSMs need to exhibit a low-power consumption typically in sub-μW regime. To achieve this power consumption, an effective way is to reduce the operating supply voltage. This reduction in voltage puts a fundamental limit on the achievable gain of the active device based amplifiers due to the reduced MOSFET stacking and reduced linearity. The subsequent sections will discuss voltage-controlled oscillator (VCO) based DSMs, inverter operational trans-conductance amplifier (OTA) based DSMs, and passive loop-filter based DSMs for realizing compact and power efficient ADCs for multichannel bio-potential acquisition systems.

4.2.1 Delta-Sigma Modulators

A $\Delta\Sigma$ modulator (DSM) operation relies on the combination of two techniques, namely: oversampling and noise shaping. For an input signal with bandwidth BW, Nyquist theorem imposes a limit on the sampling frequency, $f_s = 2 \cdot BW$. DSMs fall under the category of oversampling ADCs with an oversampling ratio (OSR) defined as $OSR = f_s/(2 \cdot B)$. The resolution of an oversampling ADC can be enhanced by filtering the in-band quantization noise such that most of the noise power lies outside the signal band, termed as noise shaping. In a DSM the noise shaping is achieved by means of negative feedback (by subtracting the input signal from an analog version of the quantizer output, and a shaping filter with an appropriate noise shaping transfer function called noise transfer function (NTF), which is usually high pass in nature. The NTF is realized using a loop filter (realized using integrators) inside the negative feedback loop. Based on the implementation type of the loop filter and hence the integrator, the DSM can be broadly divided into a continuous-time DSM (CTDSM) and a discrete-time DSM (DTDSM) as shown in Fig. 4.5. The CTDSM takes a continuous-time input and uses a continuous-time loop filter to realize the noise shaping transfer function. The DTDSM needs a front-end sample-and-hold (S/H) circuit since the loop filter is implemented using switched capacitor networks to realize the target NTF.

The building blocks needed to realize the DSM are a loop filter, a coarse ADC, and a DAC as inferred from Fig. 4.5. To design DSMs with very low-supply voltage, the coarse ADC converges to a 1-bit coarse quantizer. A 1-bit quantizer has several advantages in that the overall ADC realized using the DSM will be always monotonic unlike a SAR ADC, which needs extremely critical matching of the DAC

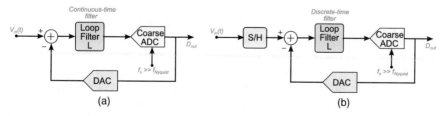

Fig. 4.5 Delta-sigma modulator implementation (**a**) with a continuous-time loop filter and (**b**) with a discrete-time loop filter

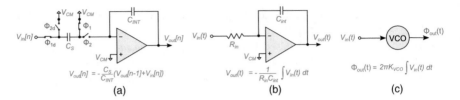

Fig. 4.6 Implementation of (**a**) DT integrator with DT voltage input and integrated DT voltage output, (**b**) CT integrator with CT voltage input and integrated CT voltage output, and (**c**) VCO based integrator with CT voltage input and CT integrated phase output

for monotonicity. With the 1-b coarse quantizer, the feedback DAC also converges to a 1-bit DAC in a DSM, which results in a simple design compared to a SAR ADC. The main building block, i.e., the loop filter, which dictates the noise shaping characteristic is implemented using integrators. A DSM with a single integrator in the loop filter exhibits a first order noise shaping of the coarse quantizer, which is 20 dB/decade. Higher order integrator realization results in better noise shaping albeit with stringent stability conditions due to the negative feedback loop.

Based on the type of nature of the loop-filter realization, the integrator can be realized using a continuous-time (CT) circuit implementation for a CTDSM or a discrete-time (DT) circuit implementation for a DTDSM. A general DT integrator is shown in Fig. 4.6a. The integrator needs two clock phases ϕ_1 and ϕ_2 for the operation. During the sampling phase ϕ_1, the input is sampled onto the capacitor C_S and during the next phase ϕ_2, the charge stored in capacitor is transferred into the integrating capacitor C_{INT}. This realizes an accumulation/integration of the input voltage over multiple clock cycles. On the contrary, the CT integrator converts the input voltage to an equivalent input current through the input resistance R_{in} and integrates the CT current on the integrating capacitor C_{int}. Assuming ideal Op-amps with infinite gain, bandwidth, and slew rate, the corresponding output voltages are given in Fig. 4.6. It must be noted that with practical Op-amps, the finite gain, bandwidth, and slew rate especially with low-voltage and sub-μW power designs, the integral realization deviates from the ideal response and limits the performance of the DSM in terms of achievable resolution. One of the main bottlenecks in the design of low-voltage and sub-μW power DSMs stems from the design of integrators with sufficient linearity, gain, bandwidth, and slew rate with a limited voltage headroom and bias current. This chapter discusses architectural realizations of the integrator for DSM design with inverter based integrator and passive integrator based designs to achieve the target power, area, and energy efficiency for the multichannel bio-potential acquisition system. Further, VCO based DSMs will also be discussed since a VCO based DSM provides a nice alternative for an integrator realization at low-supply voltages. A VCO takes the analog input voltage and the output phase of the VCO can be represented as the integral of the input voltage. To provide a fair comparison of the performance of the various DSM designs based on active integrators, passive integrators, and VCO based integrators, it is essential to define the standard figure-of-merits before indulging into the details of the DSM design details.

4.2.2 Figure-of-Merits for $\Delta\Sigma$ Modulators

The performance comparison of the $\Delta\Sigma$ modulators (DSMs) implemented with various architectures can be performed with the well-known standard figure-of-merits (FoM) defined for ADC performance evaluation. We briefly describe the FoM used to compare the DSM architectures discussed in this chapter as follows:

$$FoM_1(dB) = SNDR_{dB} + 10 \cdot log(BW/Power) \tag{4.1}$$

$$FoM_2(dB) = DR_{dB} + 10 \cdot log(BW/Power) \tag{4.2}$$

$$FoM_3(J/conv.) = Power/(2 \cdot BW \cdot 2^{ENOB}) \tag{4.3}$$

$$FoM_4 = Area/(L_{min} \cdot W_{min}) \tag{4.4}$$

FoM_1 relates the achievable signal-to-noise-distortion ratio (SNDR) in the target bandwidth and the total power consumption. A DSM with higher FoM_1 achieves better performance in terms of achieved linearity and power consumption over the target bandwidth compared to a DSM with lower FoM_1 value. FoM_2 targets the dynamic range (DR) of the DSM, which is the ratio of maximum supported amplitude and the minimum supported amplitude over the target bandwidth. Higher FoM_2 indicates that the DSM exhibits larger dynamic range for a given power consumption compared to a DSM with lower FoM_2. FoM_3 quantifies the energy efficiency of the DSM for a single-bit conversion (i.e., with unit of joules/conversion). An energy-efficient DSM exhibits a lower value of FoM_3. Finally, FoM_4 provides insight into the area efficiency of the design. To account for the designs implemented in various CMOS technology process, the total DSM design area is normalized by the minimum achievable aspect ratio in a CMOS technology. Smaller FoM_4 indicates a highly compact design. For multichannel biomedical applications, the target FoM_3 must be usually well below 100 fJ/conv. indicating a high energy efficiency and FoM_4 value well below 10 to indicate a compact design.

4.2.3 DSMs with Inverter Based Active Integrators

The main challenge in implementing integrators at low-supply voltage to realize low-voltage DSMs is the design of transductors with sufficient linearity and gain at the operating supply voltage. While transconductors can be implemented by biasing transistors in weak inversion region due to the superior gm/ID in the sub-threshold region of operation, the reduced voltage headroom and signal swings severely affect the achievable signal-to-noise ratio and linearity [19]. Inverter based operational

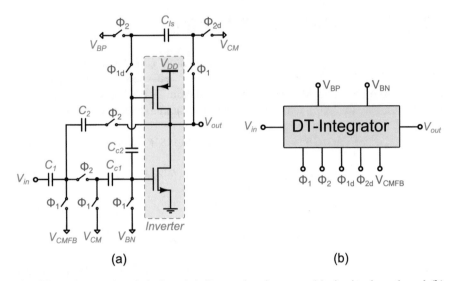

Fig. 4.7 An inverter based single-ended discrete-time integrator (**a**) circuit schematic and (**b**) equivalent block diagram

transconductance amplifier (OTA) is an efficient way of implementing low-supply voltage data converters and filters [20, 21].

Consider the inverter based active switched capacitor integrator for a low-voltage operation of 250 mV [20] as shown in Fig. 4.7a. The integrator takes the input voltage V_in, clock phases ϕ_1, ϕ_{1d}, ϕ_2, and ϕ_{2s}, and bias voltages VBP and VBN and generates the integrated output voltage V_{out} in the discrete-time domain. The integrator operated in two phases where the phase ϕ_1 corresponds to the biasing and offset compensation and ϕ_2 corresponds to the integration and amplification phase. Due to the use of only two transistors, the supply voltage can be reduced considerably. However, certain challenges must be addressed for the proper operation of the integrator circuit.

Due to the absence of bias current transistor, the bias currents in the inverter transistors are not well defined. Further, the absence of the bias current head/tail transistor severely affects the common-mode rejection property of the amplifier. To address these shortcomings, the inverter based integrator is capacitively biased, as shown in Fig. 4.7.

The operation of the inverter based integrator is illustrated in Fig. 4.8 separately in the two operational phases ϕ_1 and ϕ_2. Referring to Fig. 4.8a, during the clock phase ϕ_1, the input capacitor C_1 is charged to V_{in} referred to V_{CMFB} generated by a common-mode feedback circuit. This ensures that during the next operational phase ϕ_2, the common mode is subtracted from the input before being applied to the inverter for amplification. Capacitor C_{c1} is charged to (VBN-VCM) and the NMOS is biased at gate voltage of VBN, which is close to the NMOS threshold voltage. Capacitor C_{c2} is charged to $(V_{BP} + V_{BN})$ implying that the PMOS source-gate

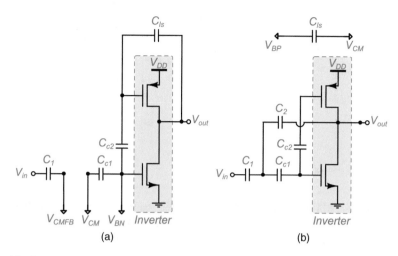

Fig. 4.8 Operation phases of the inverter based single-ended discrete-time integrator (**a**) in the bias and offset compensation phase and (**b**) in the integration phase

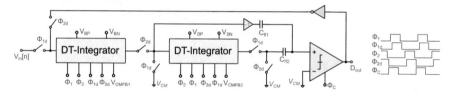

Fig. 4.9 An example single-ended second order DTDSM implemented with the inverter based switched capacitor integrator

voltage is biased at $(V_{DD} - V_{BP})$, which is set close to the threshold voltage of the PMOS transistor. This ensures that the NMOS and PMOS transistors are biased independently allowing for separate compensation of the variations in the threshold voltages [20]. Further, for offset compensation, during phase ϕ_1, the PMOS gate is connected to the inverter output through a level shifting capacitor C_{ls} (which has been pre-charge to $V_{CM} - V_{BP}$ during phase ϕ_2). This ensures that the inverter output settles to $(V_{CM} + V_{off})$ when the PMOS and NMOS area biased near their respective threshold voltages. Capacitor C_{ls} hence stores the PMOS bias voltage V_{BP} and the offset voltage V_{off}. During the phase ϕ_2, the charge on capacitor C_1 is transferred to the integrating capacitor C_2, completing the integration operation.

An example single-ended second order DTDSM implemented with the inverter based switched capacitor integrator is shown in Fig. 4.9. Each of the integrators operate at alternate clock phases and the integrator output nodes are disconnected from the integrator and are reset during the non-integration phase. The DAC is implemented with a simple inverter. The feedforward addition is implemented using capacitors Cff1 and Cff2. It must be noted that the common-mode feedback for each of the integrator must be generated from the corresponding integrator output. A 3rd

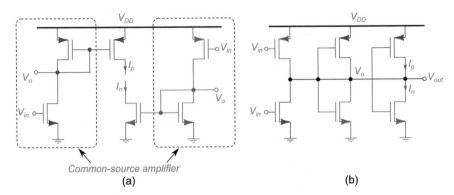

Fig. 4.10 (**a**) An NMOS and PMOS input single-ended common source amplifier with diode connected load and (**b**) inverter based amplifier for CT integrator implementation (single-ended)

order DTDSM with a supply voltage of 250 mV for a 10 kHz bandwidth was realized in [20], achieving an SNDR of 61 dB with a power consumption of 7.5 μW.

To implement a CTDSM, the integrator shown in Fig. 4.7 cannot be used due its operation being dependent on multiple clock phases. However, an active integrator with an inverter based amplifier can be realized to design a CTDSM operating at low-supply voltage. With an inverter in open loop, the output common mode is not well defined and very sensitive to transistor process variations. Figure 4.10a shows a NMOS input common source (CS) amplifier with a PMOS load and a PMOS input CS amplifier with an NMOS load. Each of the CS amplifiers have a well-defined operating point voltage V0 due to the inherent negative feedback due to the respective diode connected loads. A cascaded amplifier can be thought of with the two CS amplifiers merged as shown in Fig. 4.10b. The first stage provides an enhanced transconductance, which is the sum of NMOS and PMOS transconductance, and the second stage provides a well-defined operating point. However, it must be noted that the output common mode of the third stage is not defined and will be sensitive to PVT variations.

The modified pseudo-differential amplifier using inverters is shown in Fig. 4.11, with a cross-coupled inverter output stage. The diode connected load now provides the bias to the output stage as well solving the problem with the single ended design in Fig. 4.10b. The cross-coupled inverter introduces a regenerative feedback leading to negative conductance. The amplifier design must take into consideration the stability and PVT robustness due to the cross-coupled connection. To simplify the design, the second stage and the third stage inverters can be sized identical to avoid these issues at the cost of extra gain enhancement [21]. The inverter based pseudo-differential amplifier can be used to realize two-stage OTA with proper compensation circuit to realize the active integrator for CTDSM design. As an example, a second order CTDSM is demonstrated in Fig. 4.12. Each of the integrator is synthesized using an external RC and the active transconductor implemented using a two-stage inverter based transconductor discussed in Fig. 4.11. The feed-

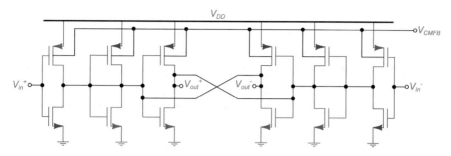

Fig. 4.11   A pseudo-differential amplifier implemented with inverters for CT integrator implementation

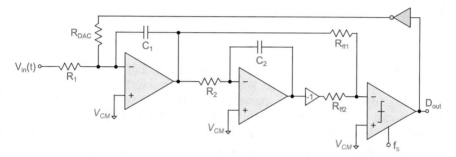

Fig. 4.12   An example second order CTDSM implemented with an inverter based active integrator

forward addition is implemented using resistors and a single-bit comparator is used with a simple inverter and resistor based 1-bit DAC to realize the CTDSM. A fourth order CTDSM is implemented using the inverter based OTA achieving a SNDR of 68 dB at a supply voltage of 300 mV [21].

While inverter based integrators for DSMs offer performance benefits in terms of reduced supply voltage, they still suffer from power consumptions that are sub-optimal for a large range of biomedical applications involving multichannel acquisition. Recently, a more power efficient way of implementing the loop filter in the DSM was introduced by realizing passive DT and CT integrators.

4.2.4   Passive Integrator Based DSMs

To completely eliminate the power consumption of the OTAs in the DSM loop-filter, passive integrator based DSMs have been explored for bio-potential acquisition and recording [17, 22–25]. In this section we discuss the CT and DT implementation of passive integrator based DSMs.

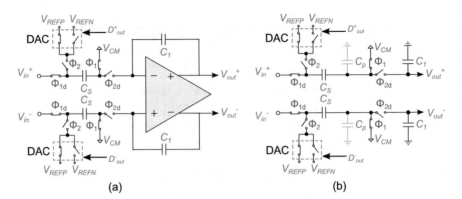

Fig. 4.13 (**a**) A differential DT active integrator and DAC subtraction and (**b**) a pseudo-differential DT passive integrator and DAC subtraction

Consider the DT active integrator with the DAC in Fig. 4.13a. The amplifier needed to implement the integrator could be based on inverter based amplifier for low-voltage operation as was discussed in the previous section or could be based on traditional OTAs for higher supply voltage operation. In either case, assuming the amplifier has a large enough gain, the integrator transfer function is given by $H(z) = \frac{C_s}{C_1} \frac{z^{-1}}{1-z^{-1}}$ meaning a pole at dc ($z = 1$). Now consider the differential DT passive integrator implementation in Fig. 4.13b. Due to the absence of any amplifier, fundamentally the passive integrator suffers from a finite dc gain. This can be understood from the passive integrator transfer function given by $H(z) = z^{-1}/((1 + C_1/C_s) - (C_1/C_s)z^{-1})$ implying that the pole at $z_p = 1/(1 + C_s/C_1)$ has moved away from dc compared to the active DT integrator. Further, the dc gain for the passive integrator has dropped to 1. If we now consider the finite parasitic capacitance Cp, in the passive integrator as shown in Fig. 4.13b, the effective transfer function is $H(z) = z^{-1}/((1 + C_1/C_s + C_p/C_s) - (C_1/C_s)z^{-1})$ leading to a dc gain of $Cs/(Cp + Cs) < 1$. This loss in dc gain and a finite pole position away from dc frequency are the two fundamental drawbacks of the passive integrator implementation.

When the DT passive integrator is used in a DTDSM for the loop-filter implementation similar to the one shown in Fig. 4.9, the quantization noise shaping is severely affected due to the attenuation in the loop and the integrator pole being away from dc frequency. Effectively, the noise transfer function of the DSM will not be aggressive compared to an active integrator implementation leading to a loss in achievable SNR but with a lower power consumption.

The problem of attenuation in the passive integrator implementation can be solved through a N-stage switched capacitor gain boosted passive integrator shown in Fig. 4.14. During the clock phase ϕ_1, all the N sampling capacitors C_s are all connected in parallel. Next, during the clock phase ϕ_2, all the N sampling capacitors C_s are reconnected in a series connection leading to an effective charge amplification on the integrating capacitor C_1 by a factor of N. Ideally, the N-

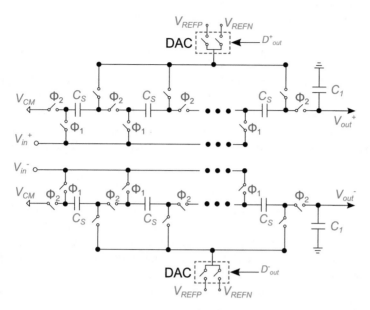

Fig. 4.14 A pseudo-differential DT passive integrator realized with a N-stage gain boosted switched capacitor network

stage gain boosting achieves a dc gain of N as against the passive integrator shown in Fig. 4.13b, which achieves an ideal dc gain of 1. This can be derived from the effective transfer function of the N-stage gain boosted integrator given by $H(z) = z^{-1}/(1/N + C_1/C_s(1 - z^{-1}))$. However, it must be noted that while ideally the achievable gain is N, this is highly limited by the parasitic capacitances in the implementation. As an example, for a 5-stage gain boosted integrator implementation, the achievable dc gain was 3.25 in a 65 nm CMOS technology [23].

The DT passive integrators discussed in this section allows for a low-voltage and low-power implementation due to the absence of active amplifier stages. However, this causes severe attenuation in the loop filter of the DSM. Hence, the required gain in the loop filter must be provided by the 1-bit comparator in the loop [17, 23]. While the signal at the comparator input is severely attenuated due to the passive integrator implementation, effectively the large comparator gain provides the sufficient loop gain for the DSM operation. A DTDSM with a second order loop filter realized with an active integrator followed by a passive integrator was realized in [23] with a 900 mV supply voltage. Also, a DTDSM with a second order loop filter with a passive integrator and a five stage gain boosted passive integrator was also demonstrated in [23]. While the DTDSM implementation with passive integrators is attractive due to the low-supply voltage and power efficiency, the design still needs a front-end sample-and-hold circuit with sufficient linearity and drive leading to overhead power consumption. Further, the multiple clock phases involved in the integrator operation must be generated on-chip leading to further power consumption overhead. To alleviate these issues CTDSM with passive integrators have been proposed for similar applications.

A passive integrator based second order CTDSM can be implemented simply with a cascade of simple passive RC integrators as shown in Fig. 4.15. Compared to a passive DTDSM implementation, it is evident that the CTDSM implementation is simpler in terms of the number of clock phases and the absence of a front-end sample-and-hold circuit. However, the feed-forward addition is implemented in the comparator itself by adding parallel transistor input paths in the comparator. The challenge in the design of the passive integrator based CTDSM is the efficient choice of component values for a target resolution in the presence of passive RC component process variations. While on-chip passive resistor and capacitor implementation with large area leads to better performance due to minimal process corner variation, it is not suitable for multichannel systems due to the large area overhead. Hence, there needs to be an optimal choice of R and C values for a target resolution, area, and power across process variations. To achieve this, a linear model for the CTDSM shown in Fig. 4.15 can be thought about for simpler optimization development for component selection. A linear model proposed in [17] is shown in Fig. 4.16, which includes the quantizer modelled as a linear gain G with an additive quantization noise e[n] modelled as white for simplicity.

A brief illustration of an optimization methodology is shown in Fig. 4.17. The optimization algorithm takes the target SNR, area, and power over the required bandwidth as inputs along with the foundry data involving the area vs. R and C for poly-resistor and MIM capacitors as input. To account for the RC process variation,

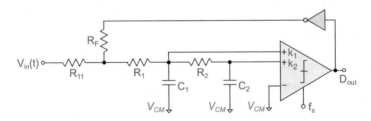

Fig. 4.15 A passive integrator based second order CTDSM implementation

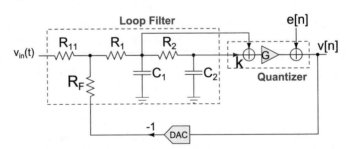

Fig. 4.16 A simplified linear model of the second order passive integrator based CTDSM

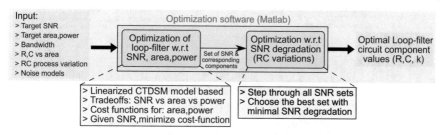

Fig. 4.17 Illustration of an optimization methodology for the passive integrator based CTDSM

the data from the foundry relating the variations in RC across different foundry defined process corners are also taken as input. Finally, the noise models also need to be considered to effectively model the achievable SNR. The optimization proceeds in two steps, first using the linearized model of the CTDSM for the target SNR a set of RC values that satisfy the minimal cost function associated with area and power are evaluated. To achieve this, the linear model is used to analytically model the various trade-offs associated with the achievable SNR, area, power, and RC values and certain cost functions are associated. Second, the set of data that satisfy the minimal cost function are processed and the one set of data that provides the minimal SNR degradation across the RC process variation while still achieving the highest SNR is chosen. The output of this optimization including the R and C values along with the feed-forward coefficient is plugged into the circuit shown in Fig. 4.15 to realize the complete CTDSM.

A second order passive integrator based CTDSM shown in Fig. 4.15 was implemented in a 180 nm CMOS technology operating at 1V supply voltage. For a target bandwidth of 10 kHz and 10-bit resolution, the optimization output yielded the resistance values of $R1 = 2\,M\Omega$, $R2 = 2\,M\Omega$, $RF = 1\,M\Omega$, $R11 = 0.9\,M\Omega$ and capacitors $C1 = 8.52\,pF$ and $C2 = 3.2\,pF$ with $k = 2$. The resistors and capacitors were realized using poly-resistors and MIM capacitors.

A post-silicon measured 64k point FFT spectrum of the passive integrator based CTDSM for a differential peak input of 700 mVpp for a near 1 kHz input is shown in Fig. 4.18. A peak SNDR of 55.42 dB and a SNR of 62.2 dB was achieved with a total power consumption of only 590 nW. The CTDSM occupies an area of only $0.068\,mm^2$. Further, no extra clock phases are required leading to no overhead power consumption in the clock generation unlike the passive integrator based DTDSM designs.

4.2.5   VCO Based DSMs

Voltage controlled oscillator (VCO) based DSMs have been increasingly popular due to the inherent first order integration property of the VCO and its mostly digital implementation [18]. The VCO output phase $\phi(t)$ is proportional to the integral of

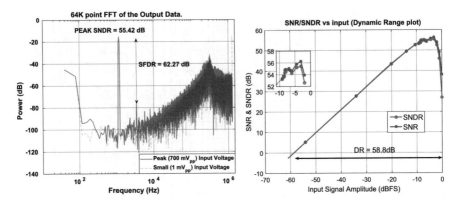

Fig. 4.18 A 64k point FFT plot of the 1-bit CTDSM output implemented with a second order passive integrator based loop filter (left) and the dynamic range plot for the CTDSM (right)

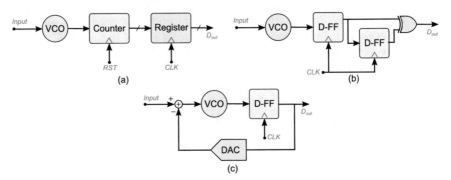

Fig. 4.19 General idea behind the VCO based DSM with (**a**) conventional counter based implementation, (**b**) an XOR differentiator based implementation, and (**c**) a DAC based closed-loop implementation

the oscillation frequency dictated by the input x(t), i.e., $\phi(t) = 2\pi \int K_{vco} x(t) dt$, where K_{vco} is the VCO tuning coefficient. The general idea behind the VCO based DSM is to utilize this first order integration of the VCO to implement the loop filter and hence the signal and noise transfer function of a DSM based ADC. A conventional VCO and counter based DSM is illustrated in Fig. 4.19a. The input to be digitized is sampled and the discrete-time analog signal is applied as the input to the VCO. The DSM is implemented in an open-loop configuration with a multi-bit counter counting the number of transitions of the VCO output in the sampling period. The output of the counter is stored in a register and the counter is reset before the start of next conversion. Briefly, the counting/accumulation provides the Σ-operation and the Δ-operation is performed by the counter reset. While the implementation is very simple, the need for a highly precise reset signal that performs the Δ-operation is one of the main bottlenecks [18]. Further, the VCO

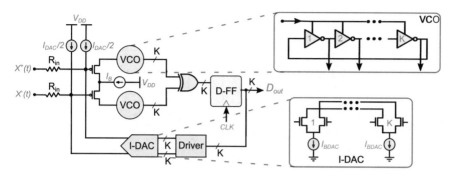

Fig. 4.20 A VCO based delta-sigma modulator with a continuous-time input to the VCO

non-linearity is a major factor limiting the achievable signal-to-noise-distortion ratio (SNDR) due to the open-loop implementation.

To avoid the use of the highly precise reset operation in the DSM shown in Fig. 4.19a, another architecture for the VCO based DSM banks on the digital implementation of (1-z-1) using a D-flip-flop and an XOR gate to realize the Δ-operation. However, the use of an XOR gate limits the dynamic range of the VCO's output phase (0 to π only) since the XOR cannot discriminate the polarity of its two inputs. Moreover, the DSM is still operated in an open-loop configuration leading to a limited achievable resolution due to the VCO non-linearity. A solution to both the problems associated with the implementation in Fig. 4.19b, i.e., the limited dynamic range and VCO non-linearity is solved with the DSM implementation shown in Fig. 4.19c. The Δ-operation is performed with a digital-to-analog converter (DAC), which forms a negative feedback loop with the VCO and the D-flip-flop. This implementation maximizes the dynamic range of the VCO's output phase from 0 to 2π. Since the VCO itself is placed inside a negative feedback loop, the non-linearity characteristic is alleviated up to certain extent leading to better achievable resolutions.

The VCO based DSM can be implemented with the VCO input being a continuous-time input or a discrete-time input. As a case study for the continuous-time input VCO based DSM, consider the schematic shown in Fig. 4.20. The DSM consists of a differential VCO and each VCO consists of a K stage current-starved ring oscillator. A differential change in the VCO input caused the phase of the differential VCO to be of opposite polarity. Comparing the phases of all the K VCO outputs using the XOR gate followed by K D flip-flops quantizes the phase output of the VCO integrator with K levels. A multi-bit current DAC takes K-bit thermometer code and provides the negative feedback in the analog domain.

The number of stages in the ring oscillator, K, which is also the number of quantization levels in the DSM loop must be chosen to keep the VCO input within its linear range. An added advantage of the architecture is the inherent dynamic element matching due to the rotation of individual DAC elements based on the thermometer code with a speed of twice that of the VCO center frequency [26]. A 300 mV supply

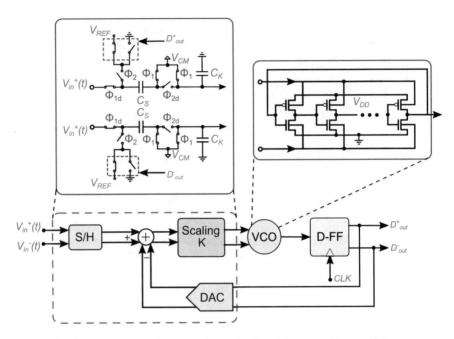

Fig. 4.21 A VCO based delta-sigma modulator with discrete-time input to the VCO

voltage DSM was demonstrated for bio-potential acquisition with a resolution of 9 bits and a power consumption of only 510 nW in a 10 kHz biomedical band [26].

As an example for the discrete-time input VCO based DSM, consider the DSM schematic diagram shown in Fig. 4.21 [18] comprising of a front-end sample-and-hold (S/H) circuit, a gain scaling block, a VCO, D flip-flop, and a 1-bit digital-to-analog converter (DAC). The input signal is sampled onto the sampling capacitor C_s during the clock phase ϕ_1. During the clock phase ϕ_2, the DAC voltage is subtracted from the sampled input voltage based on the digital output Dout, to realize the negative feedback. The clock phases ϕ_1 and ϕ_2 are complementary non-overlapping clocks and ϕ_{1d} and ϕ_{2d} are the delayed versions of the corresponding clocks. To reduce the non-linearity of the VCO, the input voltage swing of the VCO is suppressed by a factor K (attenuation factor) through a scaling block. The scaling block is also incorporated in the switched capacitor network and is decided by the factor $C_s/(C_s + C_K)$. The VCO integrates the voltage difference at its input and produces a corresponding phase output, which is then quantized by the D flip-flop. The DFF produces an output logic high if the accumulated phase exceeds 180°. The differential VCO is implemented with a bulk controlled CMOS technology for low-voltage and subsequently low-power operation [18]. A 379 nW power DSM for a 2 kHz bandwidth covering a small range of biomedical applications has been demonstrated based on the architecture in Fig. 4.21.

VCO based DSMs provide advantages in the form of compact design due to the use of largely active devices (no large passive capacitors) with a sub-μW power consumption largely due to the absence of power hungry OTAs, for biomedical applications. However, it must be noted that the VCO based DSMs provide only first order noise shaping limiting the achievable resolution. Further, the VCO non-linearity also limits the dynamic range of the ADC.

4.2.6   Performance Comparison of the DSMs

Table 4.1 compares state-of-the-art low-voltage DSMs based on the architectures discussed in this chapter in the order they were introduced. The FoMs defined earlier are used for the performance comparison of the implementations. It is evident from the table that the OTA based designs consume higher power consumption for similar performance compared to the passive integrator and VCO based integrator designs. This leads to worst energy efficiency (FoM3) for the inverter OTA based designs. However, they can operate at supply voltages as low as 250 mV. The VCO based DSMs have the best area performance (FoM4) due to the absence of passive resistors and capacitors that are absolutely necessary in the passive integrators and the inverter OTA based integrators. However, the VCO based integrators suffer from limited dynamic range owing to the non-linearity of the VCO and cannot be used for higher order noise shaping (since the inherent integration in a VCO is only first order). The passive integrator based CTDSMs offer a nice trade-off between the power consumption/energy efficiency (FoM3) and area occupancy (FoM4) between that of the inverter OTA based and the VCO based DSMs.

4.3   Direct-Digital Converters for Wearable Sensor Applications

The output signal of such signal conditioning circuits is often in the form of a change in the voltage, current, or any other analog information. An additional analog-to-digital converter is then needed to convert the voltage signal into an equivalent digital representation for further processing or decision making, which increases the complexity and power consumption of the sensor system. Therefore, direct-digital converter (DDC)—which converts the sensor's output directly into digital bits—is desirable. A DDC can be employed directly to interface resistance, capacitance, or impedance, resulting in a resistance-to-digital converter, capacitance-to-digital converter, and impedance-to-digital converters, respectively [28].

Table 4.1 Architectural and performance comparison of reported low-voltage and power/energy efficient DSMs

Parameters	[20] JSSC'12	[21] JSSC'19	[23] TCAS'14	[24] JSSC'18	[17] TCAS'20	[25] TVLSI'20	[27] ESSC'16	[18] SVLSI'13
Tech. [nm]	130	130	65	65	180	180	65	180
DSM Type	DT	CT	DT	DT	CT	CT	CT	DT
Order	3	4	2	2	2	2	1	1
Integrator-1	Inverter	Inverter	Passive	Passive	Passive	Passive	VCO	VCO
Integrator-2	Inverter	Inverter	Passive 5-SGA	Passive 5-SGA	Passive	Passive	NA	NA
Integrator-3	Inverter	Inverter	NA	NA	NA	NA	NA	NA
Integrator-4	NA	Inverter	NA	NA	NA	NA	NA	NA
Supply [V]	0.25	0.3	0.7	0.3	1	0.3	0.3	1
Clk. Freq. [MHz]	1.4	6.4	0.5	0.256	2.56	2.56	1.28	0.25
B.W [kHz]	10	50	0.5	3	10	10	10	2
OSR	70	64	500	42.67	128	128	64	62.5
Power [μW]	7.5	26.3	0.43	0.18	0.59	0.22	0.51	0.379
SNDR [dB]	61	68.5	65	60	55.42	54.81	56.1	58.5
DR [dB]	57	72.2	53	65	58.8	57.4	–	–
Area [mm^2]	0.338	0.014	0.35	0.15	0.068	0.0464	0.015	0.06
FoM$_1$ [dB]	152.2	161.3	155.6	162.2	157.7	161.38	159.02	155.7
FoM$_2$ [dB]	148	165	144	167.2	161	162.4	–	–
FoM$_3$ [fJ/conv.]	410	121.2	296	36.7	61.1	24.47	49	137.8
FoM$_4$	20	0.83	72.91	31.14	1.57	1.07	3.55	1.38

4.3.1 Resistance-to-Digital Converter

The resistance-to-digital converters are used to convert the sensor resistance into equivalent digital bit-streams. The resistive sensor in such converters is part of the direct-digital conversion. In this section, we discuss a range of resistance-to-digital converters suitable for wearable sensor applications.

4.3.1.1 Microcontroller Based Direct-Digital Converter for Resistive Sensor

The circuit diagram of the direct microcontroller based resistance-to-digital converter, reported in [27], is shown in Fig. 4.22. The RDC is based on charging/discharging of a reference capacitor. The sensor resistance is shown by Rx and the reference capacitor is represented by C_i. Both resistance and capacitance are connected to the pins of the microcontroller as shown in Fig. 4.22a. The capacitor C_i charges till the voltage of the capacitor reaches the V_{DD} as shown in Fig. 4.22b. After a fixed charging time period, the capacitor Ci discharges through the sensor resistance Rx. The number of clock pulses needed to discharge the capacitor to a fixed threshold voltage is proportional to the sensor resistance.

The time needed to discharge the capacitor voltage through the resistor (as shown in Fig. 4.22) can be written as follows:

$$T = R_x C_i \frac{V_{DD} - V_{ss}}{V_{TL} - V_{ss}} \tag{4.5}$$

The time period is converted into digital counts by counting the number of clock pulses in T using the microcontroller's timer module. The improved direct-digital converters are reported in [29].These improved direct-digital converters provide a

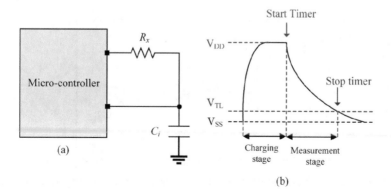

(a)

(b)

Fig. 4.22 The schematic diagram of the direct sensor to microcontroller interface

linear change in the period T regarding the sensor resistance. Moreover, the direct-digital converter circuits are also used for full-bridge based resistive sensors and the differential resistive sensors [30, 31]. The microcontroller based direct-digital converters are considered to be the simplest digital converters for the sensors.

4.3.1.2 Incremental Delta-Sigma Based Direct-Digital Converter for Resistive Sensors

The delta-sigma based RDC, utilizing oversampling and noise shaping, is widely used in high-accuracy measurement applications [30]. The delta-sigma based RDC is designed to deliver good sample-by-sample conversion performance. In addition, the incremental delta-sigma converters exhibit excellent differential and integral linearity, low offset and gain errors, and high resolution and low-power consumption [31].

The schematic diagram of a first order delta-sigma converter for resistive sensors is shown in Fig. 4.23. The voltage from the sensor branch Rx is continuously integrated and compared with the common-mode potential using the ZCD. The reference signal, from the reference branch Rref, is subtracted from the sensors' branch when the integrator's output cross the common-mode potential. Once it reaches the common-mode potential, the output voltage of the ZCD will flip to the opposite polarity (from high to low). The next rising edge of the clock signal changes the D flip-flop's output logic level, which subsequently switch $-V_i$ to R_{ref}. The negative polarity of $-V_i$ is now added with the input signal V_i, which results in a sharp transition in the integrator output voltage. The output of ZCD becomes high again. This subsequently changes the D flip-flop's output logic level. The process is repeatable, and we obtain a bit-stream at the output of D flip-flop. This form of charge balancing ensures that the average charge accumulated at the integrator's output is approximately zero. The information about the sensor parameters can be obtained by counting the total numbers of clocks [30, 31].

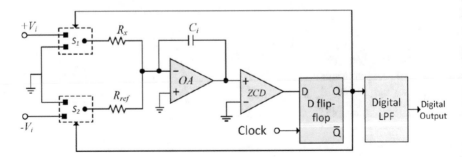

Fig. 4.23 Schematic diagram of the incremental delta-sigma based resistance-to-digital converter

4.3.2 Capacitance-to-Digital Converters

Flexible capacitive sensors are used in energy-constrained applications, as they do not consume static power and can be used in a wide range of applications to measure different physical, chemical, or biological quantities. The capacitance-to-digital converter (CDC) is used to convert the output of the sensor into equivalent digital bits. In this section we will discuss some of the useful CDC topologies.

4.3.2.1 Dual-Slope Based Capacitance-to-Digital Converter

The dual-slope based analog-to-digital converters are widely used to convert the analog voltage of the sensors—such as thermocouple and bio-potential signal—into equivalent digital bits. The dual-slope analog-to-digital converters comprise two integration cycles for the measurement: (i) integration of unknown analog input signal and (ii) integration of a known reference signal. The time taken by both measurement cycles are converted into equivalent digital bit-streams. The obtained bit-streams are proportional to the unknown analog input voltage [30, 32, 33].

Usually for sensor interfacing, the sensor capacitance is first converted into equivalent voltage signal and an analog-to-digital converter is utilized to convert the proportional voltage signal into equivalent bit-stream. This is a two-step process, which increases the power consumption and complexity of the sensor system.

Instead of converting the sensor capacitance into voltage signal, the dual-slope based direct digitizer is configured by using the sensor capacitor as part of the converter. This way the output bit-stream is proportional to sensor capacitance directly. This enables the possibility of designing low-power capacitance-to-digital converters for applications such as wearable and implantable [34–37].

A dual-slope based integrated capacitance-to-digital converter for capacitive sensors is shown in Fig. 4.24. The operation of the circuit is based on transferring charge from the sensor capacitance to an integrating capacitor. The sensor capacitance is shown as C_{sen}. A capacitor C_{base} is included to compensate the baseline capacitance of the sensor. The charge Q_e proportional to the difference between C_{sen} and C_{base} flows through the integrator capacitor C_{int}.

The operation of the circuit is divided into two phases: (i) the sampling phase and (ii) the discharge phase. During the sampling phase, the integrator capacitor is charged with the charge transferred from the sensor capacitance. The charges from the sensor capacitor are transferred in packets controlled with a clock ϕ_{s2}. Once the integrator capacitor is charged for a fixed number of clock cycles, the discharge phase begins by disabling the operational transconductance amplifier OTA$_1$ & OTA$_2$ and enabling the OTA$_3$. The value of discharge capacitor C_{ref} is smaller than the effective sensor capacitance. Each clock pulses discharge C_{int} till it reaches the threshold voltage of the comparator. The number of clock pulses required to discharge the C_{int} is proportional to the sensor capacitance.

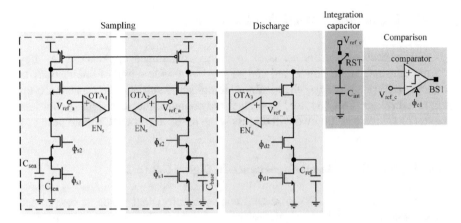

Fig. 4.24 Circuit diagram of the dual-slope based capacitance-to-digital converter for capacitive sensors

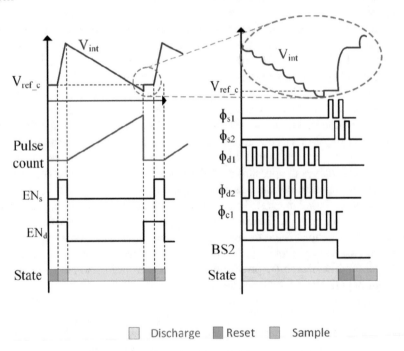

Fig. 4.25 Timing waveform of the dual-slope based CDC shown in Fig. 4.24

The waveform of the capacitance-to-digital converter is shown in Fig. 4.25. During the sampling phase, the charge transferred from the sampling capacitors to the integrating capacitor for each ϕ_s cycle can be given as follows:

$$Q_{add} = (C_{sen} - C_{base}) \times V_{ref_a} \tag{4.6}$$

For N_1 cycles, the total charge transferred to C_{int} is equal to $N_1 Q_{add}$.

Next, during the discharge state, the comparator is turned ON. The amount of charge Q_{sub} that is subtracted from C_{int} for each ϕ_c cycle and the value of voltage V_{int} at the end of the Nth cycle of the discharge stage are given as follows:

$$Q_{sub} = C_{C_{ref}} \times V_{ref_a} \tag{4.7}$$

$$V_{int}(n) = V_{ref_c} + \frac{N_1 \times Q_{add} - N \times Q_{sub}}{C_{int}} \tag{4.8}$$

The discharge phase ends when the voltage V_{int} becomes less than the threshold voltage $V_{(ref_c)}$ of the comparator. The N number of clock cycles for the integrator voltage to reach till $V_{(ref_c)}$ can be written as follows:

$$N = \frac{N_1(C_{sen} - C_{base})}{C_{ref}} \tag{4.9}$$

The number of clocks N is proportional to the sensor capacitance and can be calculated by a ripple counter. The power consumption of the circuit is reduced by disabling the OTA in the phased manner. For example, during the sampling phase, the OTA utilized in the discharge state and comparator are disabled. Similarly, during the discharge phase, the OTA utilized in the sampling phase are disabled. The resolution of the circuit can be enhanced by utilizing a course and a fine comparator as reported in [36].

4.3.2.2 Direct-Digital Converter for Leaky Capacitive Sensors

The aforementioned capacitance-to-digital converters are designed based on the assumption that the capacitive sensor is pure in nature and can be represented in equivalent electrical form by a capacitor C_x. However, most capacitive sensors, especially those with planner electrode geometry, are leaky, allowing the flow of continuous static current through the sensor. This leakage can be represented in equivalent electrical form by a parallel resistor Rx with the sensor capacitor C_x. The leakage resistance is sensitive to the surrounding environment condition, electrodes geometry, measurand type, and the fabrication process. This leakage may affect the accuracy and/or linearity of the sensor system. Therefore, the leakage resistor's effect on the measurement of the sensor capacitance needs to be compensated [38–41].

A dual-slope based capacitance-to-digital converter for leaky capacitive sensors is shown in Fig. 4.26. The converter is based on the phase-sensitive-detection (PSD) and dual-slope technique. The phase-sensitive technique using a quadrature phase shifted signal separates the capacitive components from the complex sensor impedance. Two branches are used for dual-slope operation: (i) the sensor branch and (ii) the reference branch.

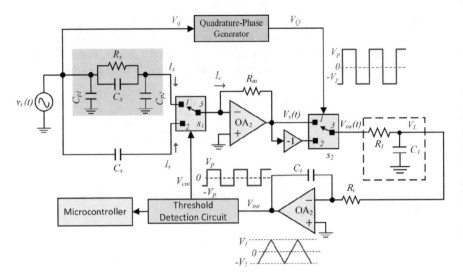

Fig. 4.26 Dual-slope based capacitance-to-digital converter for leaky capacitive sensors

The sensor branch consists of the leaky capacitive sensor and the reference branch consists of the capacitor C_s. A switch S_1 is used, which allows the selection of either sensor of reference branch. First, assume the sensor branch is connected to the input of switch S_1, and the current I_x flows through the resistor R_m and generates the voltage $V_x(t)$. Voltage $V_x(t)$ and inverted version of $V_x(t)$ are used for the phase-sensitive-detection using switch S_2. The output $V_{sw}(t)$ of the switch is low-pass filtered. The combined operation of the PSD and low-pass filter mitigates the effect of the leakage resistor. The filtered voltage V_L is integrated using an integrator. The output of the integrator is applied to a threshold detection circuit. The threshold detection circuit has a positive threshold voltage and a negative threshold voltage of same amplitude. The output of the threshold detection circuit is a square-wave signal that controls the position of switch S_1. The number of pulses needed when the switch S_1 is in position 1 is proportional to the sensor capacitance.

Phase-1 (Switch S_1 in Position-1)

$$V_{S2}(t) = \begin{cases} -V_m R_m \left[G_x \sin(\omega t) + \omega (C_x) \cos(\omega t) \right] & ; \pi/2 \; < \omega t \; < 3\pi/2 \\\\ V_m R_m \left[G_x \sin(\omega t) + \omega (C_x) \cos(\omega t) \right] & ; 0 < \omega t \; < \pi/2 \&; 3\pi/2 < \omega t \; < 2\pi \end{cases}$$

$$(4.10)$$

The output voltage of the low-pass filter can be written as follows:

$$V_{L(S_1-1)} = -\frac{2V_m R_m \omega C_x}{\pi} \tag{4.11}$$

The voltage $V_{L_{(s_1-1)}}$ is integrated by the integrator. Once the output of the integrator reaches a threshold voltage, the threshold detection circuit toggles the position of the switch S_1. A counter is set to count the number of clock pulses N_1 required for the integrator output to reach the desired threshold voltage. Assume $V_{(TH_1)}$ is the threshold voltage of the threshold detection circuit, and the output of integrator after N_1 clock cycles can be derived as follows:

$$|V_{TH_1}| = \frac{N_1 T_{clk}}{R_i C_i} \frac{2V_m R_m \omega C_x}{\pi} \tag{4.12}$$

Phase-2 (Switch S_1 in Position-2)
Similar to the phase-1, the threshold voltage $V_{(TH_2)}$ can be derived as follows:

$$|V_{TH_1}| = \frac{N_2 T_{clk}}{R_i C_i} \frac{2V_m R_m \omega C_x}{\pi} \tag{4.13}$$

Suppose the threshold voltage $|V_{(TH_1)} = V_{(TH_2)}|$, the expression of the sensor capacitor C_x can be written as follows:

$$C_x = \frac{N_2}{N_1} C_s \tag{4.14}$$

4.3.3   Impedance-to-Digital Converters

Bioimpedance measurement is a method of measurement of electrical impedance of the body to determine various body compositions such as the body fat in relation to body mass and various diseases such as the pulmonary artery diseases. The measurement of bioimpedance is commonly based on four electrode method in which two electrodes are used to provide a sinusoidal current of known amplitude and another two electrodes are used to detect the change in the bio-potential due to body impedance [9].

The conventional bioimpedance based system needs several active components such as the instrumentation amplifier (INA), filter, analog multiplier, and an analog-to-digital converter to convert the bioimpedance into equivalent digital bit-stream. The commercial integrated circuit for bioimpedance measurement such as AD5933 consumes around 130 mW power. The high-power consumption of the conventional bioimpedance measurement ICs makes them unsuitable for the wearable applications such as the electronic skin [42, 43].

4.3.3.1 Dual-Slope Based Impedance-to-Digital Converter

The signal conditioning circuits for measuring both R_x and C_x of the impedance R–C sensors are mostly based on the separation of in-phase and quadrature components [42, 44]. The schematic diagram of a simple impedance-to-digital converter based on the separation of in-phase and quadrature components from the sensors' output signal is shown in Fig. 4.27 [45, 46]. The circuit is based on converting the sensor current into equivalent voltage and demodulating the voltage using in-phase and quadrature control signals. The in-phase components of the output are proportional to the sensor resistance, and the quadrature components are proportional to the sensor capacitance. A low-pass filter is used to remove the high-frequency components from the output signals. A switch is used for the phase-sensitive-detection. The switches are controlled by the square-wave in-phase and quadrature control signals. A mode select signal was used, which sequentially measures the resistance and capacitance of the sensor as follows:

$$V_{MS} = \begin{cases} \text{Low } (S_1, S_2: \text{position-2}) \to \text{Mode C; Capacitive Sensors} \\ \text{High } (S_1, S_2: \text{position-1}) \to \text{Mode R; Resistive Sensors} \\ \text{Square Wave} \to \text{Mode-Z; Impedance Sensors} \end{cases} \quad (4.15)$$

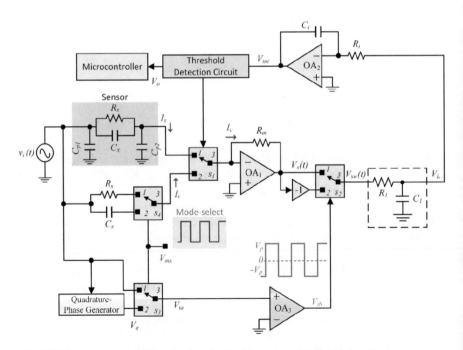

Fig. 4.27 Dual-slope based impedance-to-digital converter for parallel R–C sensors

Mode-R: Resistance Measurement

In mode-R, the switches S_3 and S_4 are in position-1. Each mode operates in two phases similar to the circuit in leaky capacitive sensors shown in Fig. 4.26. In phase-1, the switch S_1 connects the sensor to the input of OA_1. The output of the operational amplifier OA_1 is applied to switch S_3 for the phase-sensitive-detection with in-phase signal. The output of the switch is filtered and applied to the integrator and threshold detection circuitry. Once the threshold is reached, the position of switch S_1 changes to 2 and the reference resistor R_s connects at the input of OA_1. In phase-2, the voltage V_L proportional to the reference resistance integrates till it reaches the same amplitude threshold level. Considering N_1 are the number of clock cycles for the sensor resistance and N_2 are the number of clock cycles for the reference resistance, the value of the sensor resistance can be written as follows:

$$R_x = \frac{N_1}{N_2} R_s \qquad (4.16)$$

Mode-R: Capacitance Measurement

In mode-C of the operation, the switches S_3 and S_4 are in position-2. The reference capacitor C_s is used in the phase-2 of the operation of the circuit. The quadrature phase shifted signal was used for the phase-sensitive-detection function. The sensor current I_x is integrated in phase-1 of the operation and the reference current I_s is integrated in phase-2 of the operation. The rest of the operating principle of the circuit is similar to the circuit shown in Fig. 4.26 for the capacitance measurement of leaky capacitive sensors.

4.3.3.2 Delta-Sigma Based Bioimpedance-to-Digital Converter

The schematic diagram of a delta-sigma based bioimpedance-to-digital converter circuit is shown in Fig. 4.28. The loop of the delta-sigma converter consists of an integrator, a comparator, a pulse shaping circuit, and a digital-to-analog converter (DAC). The unknown impedance is connected at the input of the integrator. A current I_x flows through the integrator. The current through the DAC is given as I_r. The error current I_e integrates using the integrator. The output of the integrator is passed through a zero-crossing detector.

In the delta-sigma converters high signal-to-noise ratio (SNR) can be with high over-sampling ratio (OSR). Thanks to the narrow frequency bandwidth of the input signals, the dynamic power of the ADC can be kept low, despite the high OSR value. The noise shaping property of ADCs results in modulating the noise from low frequency to higher frequencies and the quantization noise to be much lower than Nyquist-rate ADCs at low frequencies (baseband).

The bit-stream from the output of the delta-sigma loop is demodulated using an XOR based phase-sensitive-detection approach. The in-phase and quadrature part from the output of the impedance-to-digital converter is obtained using an XOR based phase detection technique. The reference in-phase and quadrature signals are

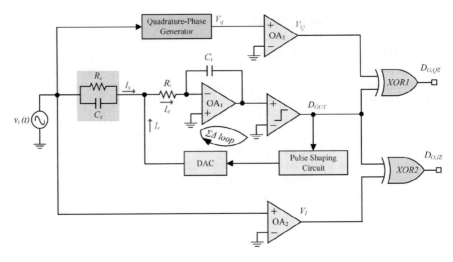

Fig. 4.28 Delta-sigma based impedance-to-digital converter

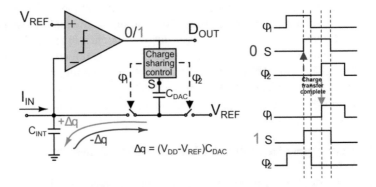

Fig. 4.29 Schematic diagram of low-power bioimpedance-to-digital converter reported in [22]

converted into the square-wave signal using the zero-crossing detectors. The zero-crossing detector is implemented using OA_2 and OA_3. The real part (R_x) of the sensor impedance is obtained by separating the bit-stream using XOR operation from the delta-sigma loop with an in-phase signal followed by a zero-crossing-detector. Similarly, the quadrature part (C_x) of the sensor is obtained by the XOR operation with the square-wave quadrature phase shifted signal.

An ultra-low-power opamp-less DSM for bioimpedance spectrum analyzer was proposed in [22]. The current I_x is from the impedance sensor. The reported work relied on a simple capacitor based first order integration of the incoming current signal. A charge-sharing DAC was used to push/pull current through a small capacitor CDAC to/from the larger integrating capacitor C1, depending on the comparator output, as shown in Fig. 4.29. The utilized comparator and the DAC complete a negative feedback loop and maintain the input node voltage at VREF. To maintain the digital negative feedback loop (virtual node at the integrating capacitor node), a relatively high clock frequency (hence OSR) is needed for the clocked comparator. Further, the input current is integrated directly by the integrating capacitor, leading to a smaller voltage range (limited DR). Briefly, while the design achieves an SNDR of 50.3 dB with a power of only 50 nW over a 5 kHz bandwidth, it has a minimum dynamic range. Further, to increase the SNDR, the OSR has to be increased significantly, leading to very high-power consumption at higher resolutions.

References

1. S. Malik, L. Somappa, M. Ahmad, S. Sonkusale, M.S. Baghini, A fringing field based screen-printed flexible capacitive moisture and water level sensor, in *2020 IEEE International Conference on Flexible and Printable Sensors and Systems (FLEPS)* (2020), pp. 1–4
2. M. Ahmad, S. Malik, L. Somappa, S. Sonkusale, M.S. Baghini, A flexible dry ECG patch for heart rate variability monitoring, in *2020 IEEE International Conference on Flexible and Printable Sensors and Systems (FLEPS)* (2020), pp. 1–4
3. L. Somappa, S. Malik, M. Ahmad, K.M. Ehshan, A.M. Shaikh, K.M. Anas, S. Sonkusale, M.S. Baghini, A 3D printed robotic finger with embedded tactile pressure and strain sensor, in *2020 IEEE International Conference on Flexible and Printable Sensors and Systems (FLEPS)* (2020), pp. 1–4
4. M. Ahmad, S. Malik, S. Dewan, A.K. Bose, D. Maddipatla, B.B. Narakathu, M.Z. Atashbar, M.S. Baghini, An auto-calibrated resistive measurement system with low noise instrumentation ASIC. IEEE J. Solid State Circuits **55**(11), 3036–3050 (2020)
5. S. Malik, M. Ahmad, L. S, T. Islam, M.S. Baghini, Impedance-to-time converter circuit for leaky capacitive sensors with small offset capacitance. IEEE Sensors Lett **3**(7), 1–4 (2019)
6. K. Kishore, S. Malik, M.S. Baghini, S.A. Akbar, A dual-differential subtractor-based auto-nulling signal conditioning circuit for wide-range resistive sensors. IEEE Sensors J. **20**(6), 3047–3056 (2020)
7. S. Malik, Q. Castellví, L. Becerra-Fajardo, M. Tudela-Pi, A. García-Moreno, M.S. Baghini, A. Ivorra, Injectable sensors based on passive rectification of volume-conducted currents. IEEE Trans. Biomed. Circuits Syst. **14**(4), 867–878 (2020)
8. S. Malik, M. Ahmad, M. Punjiya, A. Sadeqi, M.S. Baghini, S. Sonkusale, Respiration monitoring using a flexible paper-based capacitive sensor, in *2018 IEEE SENSORS* (2018), pp. 1–4
9. J.G. Webster, *Medical Instrumentation: Application and Design* (John Wiley & Sons, Hoboken, 2009)
10. I.H. Stevenson, K.P. Kording, How advances in neural recording affect data analysis. Nat. Neurosci. **14**(2), 139–142 (2011). https://doi.org/10.1038/nn.2731
11. L. Somappa, M.S. Baghini, A 400 mV 160 nW/Ch compact energy efficient modulator for multichannel biopotential signal acquisition system. IEEE Trans. Biomed. Circuits Syst. 1–1 (2021)
12. S.J. Kim, S.H. Han, J.H. Cha, L. Liu, L. Yao, Y. Gao, M. Je, A sub-μw/ch analog front-end for δ-neural recording with spike-driven data compression. IEEE Trans. Biomed. Circuits Syst **13**(1), 1–14 (2019)
13. X. Zou, L. Liu, J.H. Cheong, L. Yao, P. Li, M. Cheng, W.L. Goh, R. Rajkumar, G.S. Dawe, K. Cheng, M. Je, A 100-channel 1-mw implantable neural recording IC. IEEE Trans. Circuits Syst. Regul. Pap. **60**(10), 2584–2596 (2013)

14. A. Rodríguez-Pérez, M. Delgado-Restituto, F. Medeiro, A 515 nw, 0–18 dB programmable gain analog-to-digital converter for in-channel neural recording interfaces. IEEE Trans. Biomed. Circuits Syst. **8**(3), 358–370 (2014)
15. M.S. Chae, Z. Yang, M.R. Yuce, L. Hoang, W. Liu, A 128-channel 6 mw wireless neural recording ic with spike feature extraction and UWB transmitter. IEEE Trans. Neural Syst. Rehabil. Eng. **17**(4), 312–321 (2009)
16. M.S. Chae, W. Liu, M. Sivaprakasam, Design optimization for integrated neural recording systems. IEEE J. Solid State Circuits **43**(9), 1931–1939 (2008)
17. L. Somappa, M.S. Baghini, A compact fully passive loop filter-based continuous time $\Delta\Sigma$ modulator for multi-channel biomedical applications. IEEE Trans. Circuits Syst. I Regul. Pap. **67**(2), 590–599 (2020)
18. Y. Chang, C. Weng, T. Lin, C. Wang, A 379nW 58.5dB SNDR VCO-based $\Delta\Sigma$ modulator for bio-potential monitoring, in *2013 Symposium on VLSI Circuits* (2013), pp. C66–C67
19. A. Wang, B.H. Calhoun, A.P. Chandrakasan, *Sub-threshold Design for Ultra Low-Power Systems*, vol. 95 (Springer, Berlin, 2006)
20. F. Michel, M.S.J. Steyaert, A 250 mv 7.5 μw 61 dB SNDR SC $\Delta\Sigma$ modulator using near-threshold-voltage-biased inverter amplifiers in 130 nm CMOS. IEEE J. Solid State Circuits **47**(3), 709–721 (2012)
21. L. Lv, X. Zhou, Z. Qiao, Q. Li, Inverter-based subthreshold amplifier techniques and their application in 0.3-V $\Delta\Sigma$ -Modulators. IEEE J. Solid State Circuits **54**, 1436–1445 (2019)
22. M. ElAnsary, N. Soltani, H. Kassiri, R. Machado, S. Dufour, P.L. Carlen, M. Thompson, R. Genov, 50nW opamp-less $\Delta\Sigma$-modulated bioimpedance spectrum analyzer for electrochemical brain interfacing. IEEE J. Solid State Circuits **55**(7), 1971–1983 (2020)
23. A.F. Yeknami, F. Qazi, A. Alvandpour, Low-power DT $\Delta\Sigma$ modulators using SC passive filters in 65 nm CMOS. IEEE Trans. Circuits Syst. Regul. Pap. **61**(2), 358–370 (2014)
24. A.F. Yeknami, X. Wang, I. Jeerapan, S. Imani, A. Nikoofard, J. Wang, P.P. Mercier, A 0.3-V CMOS biofuel-cell-powered wireless glucose/lactate biosensing system. IEEE J. Solid State Circuits **53**(11), 3126–3139 (2018)
25. L. Somappa, M.S. Baghini, A 300-mV auto shutdown comparator-based continuous time $\Delta\Sigma$ modulator. IEEE Trans. Very Large Scale Integr. VLSI Syst. **28**(8), 1920–1924 (2020)
26. L. Somappa, M.S. Baghini, A sub-μW power 10 bit $\Delta\Sigma$ ADC for biomedical applications, in *2020 IEEE 63rd International Midwest Symposium on Circuits and Systems (MWSCAS)* (2020), pp. 679–682
27. N. Narasimman, T.T. Kim, A 0.3 V, 49 fJ/conv.-step VCO-based delta sigma modulator with self-compensated current reference for variation tolerance, in *ESSCIRC Conference 2016: 42nd European Solid-State Circuits Conference* (2016), pp. 237–240
28. S. Malik, L. Somappa, M. Ahmad, T. Islam, M.S. Baghini, An accurate digital converter for lossy capacitive sensors. Sens. Actuators, A **331**, 112958 (2021)
29. F. Reverter, J. Jordana, M. Gasulla, R. Pallàs-Areny, Accuracy and resolution of direct resistive sensor-to-microcontroller interfaces. Sens. Actuators, A **121**(1), 78–87 (2005)
30. E. Sifuentes, O. Casas, F. Reverter, R. Pallas-Areny, Improved direct interface circuit for resistive full-and half-bridge sensors, in *2007 4th International Conference on Electrical and Electronics Engineering* (IEEE, Piscataway, 2007), pp. 197–200
31. F. Reverter, O. Casas, Interfacing differential resistive sensors to microcontrollers: a direct approach. IEEE Trans. Instrum. Meas. **58**(10), 3405–3410 (2009)
32. E. Sifuentes, O. Casas, F. Reverter, R. Pallas-Areny, Direct interface circuit to linearise resistive sensor bridges. Sens. Actuators, A **147**(1), 210–215 (2008)
33. N.M. Mohan, B. George, V.J. Kumar, A novel dual-slope resistance-to-digital converter. IEEE Trans. Instrum. Meas. **59**(5), 1013–1018 (2010)
34. N.M. Mohan, B. George, V.J. Kumar, A sigma-delta resistance to digital converter suitable for differential resistive sensors, in *2008 IEEE Instrumentation and Measurement Technology Conference* (IEEE, Piscataway, 2008), pp. 1159–1161
35. B. George, V.J. Kumar, Analysis of the switched-capacitor dual-slope capacitance-to-digital converter. IEEE Trans. Instrum. Meas. **59**(5), 997–1006 (2010)

36. N.M. Mohan, B. George, V.J. Kumar, Dual slope resistance to digital converter, in *2007 IEEE Instrumentation & Measurement Technology Conference IMTC 2007* (IEEE, Piscataway, 2007), pp. 1–5
37. L. Somappa, S. Malik, S. Aeron, S. Sonkusale, M.S. Baghini, High resolution frequency measurement techniques for relaxation oscillator based capacitive sensors. IEEE Sensors J. **21**(12), 13 394–13 404 (2021)
38. B. George, V.J. Kumar, Analysis of the switched-capacitor dual-slope capacitance-to-digital converter. IEEE Trans. Instrum. Meas. **59**(5), 997–1006 (2010)
39. C. Rogi, C. Buffa, N. De Milleri, R. Gaggl, E. Prefasi, A fully-differential switched-capacitor dual-slope capacitance-to-digital converter (CDC) for a capacitive pressure sensor. Sensors **19**(17), 3673 (2019)
40. S. Oh, Y. Lee, J. Wang, Z. Foo, Y. Kim, W. Jung, Z. Li, D. Blaauw, D. Sylvester, A dual-slope capacitance-to-digital converter integrated in an implantable pressure-sensing system. IEEE J. Solid State Circuits **50**(7), 1581–1591 (2015)
41. S. Malik, L. Somappa, M. Ahmad, M.S. Baghini, An-c2v: an auto-nulling bridge-based signal conditioning circuit for leaky capacitive sensors. IEEE Sensors J. **20**(12), 6432–6440 (2020)
42. S. Malik, K. Kishore, T. Islam, Z.H. Zargar, S. Akbar, A time domain bridge-based impedance measurement technique for wide-range lossy capacitive sensors. Sens. Actuators, A **234**, 248–262 (2015)
43. F. Reverter, Ò. Casas, A microcontroller-based interface circuit for lossy capacitive sensors. Meas. Sci. Technol. **21**(6), 065203 (2010)
44. P. Vooka, B. George, An improved capacitance-to-digital converter for leaky capacitive sensors. IEEE Sensors J. **15**(11), 6238–6247 (2015)
45. S. Malik, M. Ahmad, L. Somappa, T. Islam, M.S. Baghini, An-z2v: autonulling-based multimode signal conditioning circuit for RC sensors. IEEE Trans. Instrum. Meas. **69**(11), 8763–8772 (2020)
46. P. Vooka, B. George, A direct digital readout circuit for impedance sensors. IEEE Trans. Instrum. Meas. **64**(4), 902–912 (2014)

Chapter 5
Power Management Circuits for Energy Harvesting

5.1 Introduction to Energy Harvesting from Ambient

Wireless sensor nodes (WSNs) or body area network plays an important role in the development of a smart world. WSNs can be used in places like environmental monitoring, surveillance, smart building, and sensors located in hazardous environments [1–9]. Frequent replacement of battery in wireless sensor nodes increases the operation cost and can be cumbersome if 100s of wireless sensor nodes are involved [10]. Therefore, with the help of the energy harvesting, operation cost and hard-work for replacing the battery in wireless sensor nodes can be either completely avoided or can be decreased. A typical wireless sensor node is shown in Fig. 5.1. A battery-less wireless sensor node consists of an energy harvester, a power management circuit to process the harvested energy efficiently, a sensor to sense the external parameters like humidity, temperature, air quality, and an application circuit powered by the power management circuit. Application circuit either receives the signals from the base station or processes the data gathered by the sensor and transmits it to base station [11–17].

Different types of energy harvester are available to harvest the energy from the ambient and power the WSN. RF energy harvesters are reported in [11, 13, 16, 25], photovoltaic energy harvesters are reported in [14, 26–30], thermo-electric energy harvesters are reported in [14, 26, 31], and piezo-electric energy harvesters are reported in [32–36]. The output power of different energy harvesters under certain ambient conditions is mentioned in Table 5.1. For very weak ambient conditions, the output of the energy harvester is few 100s of mV or even less. For thermal energy harvester of $50\,cm^2$ size, with a temperature difference of $2\,K$, the output voltage is $50\,mV$ [34]. Also, the output power is limited to few μW's or even

Dr. Maryam Shojaei Baghini (IIT Bombay), Dr. Meraj Ahmad (IIT Bombay), Dr. Shahid Malik (IIT Delhi) and Dr. Gaurav Saini (IIT Bombay) contributed to this chapter.

© The Author(s), under exclusive license to Springer Nature Switzerland AG 2022   121
S. Sonkusale et al., *Flexible Bioelectronics with Power Autonomous Sensing and Data Analytics*, https://doi.org/10.1007/978-3-030-98538-7_5

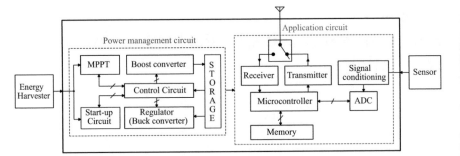

Fig. 5.1 A typical wireless sensor node comprises energy harvester, power management circuit, sensor, and application circuit. An EHS includes energy harvester and power management circuit

Table 5.1 Different types of sources and their output power density

Reference	Source	Source characteristics
[18]	RF	GSM 900 and GSM 1800
		At a distance of 25 m, available RF power density $= 100\,\text{nW/cm}^2$
		At a distance of 100 m, available RF power density $= 10\,\text{nW/cm}^2$
[19]	RF	Incident RF power $= 0.1\,\text{mW/cm}^2 = -10\,\text{dBm}$, Rectenna output power $= 0.02\,\text{mW/cm}^2 = -17\,\text{dBm}$ (frequency $= 3\,\text{GHz}$)
[18]	Vibrational	Area $= 10.75\,\text{mm}^2$, acceleration $= 0.25\,g$, frequency $= 68\,\text{Hz}$, mass $= 0.49\,\text{mg}$, output power $= 23\,\text{nW}$
[20]	Vibrational	Area $= 2.65\,\text{mm}^2$, acceleration $= 1\,g$, frequency $= 608\,\text{Hz}$, output power $= 2.16\,\mu\text{W}$
[21]	Thermal	Area $= 9\,\text{cm}^2$, $\Delta T = 0.5\,°\text{C}$, output power $= 20\,\mu\text{W}$, internal resistance $= 7.6\,\Omega$, power density $= 2.2\,\mu\,\text{W/cm}^2$
[22]	Thermal	Area $= 9\,\text{cm}^2$, $\Delta T =$ difference between human body and room temperature, output power $= 100\,\mu\,\text{W/cm}^2$
[23]	Photovoltaic cell	Device is InGaP-GaAs. Area $= 1\,\text{cm}^2$, $V_{oc} = 2.547\,\text{V}$, $I_{sc} = 14.3\,\text{mA/cm}^2$, Maximum power $= 30.79\,\text{mW/cm}^2$ at 1 Sun intensity
[24]	Photovoltaic cell	GaInP cell, Area $= 45\,\text{cm}^2$, output power $= 15.6\,\mu\text{W/cm}^2$ at 200 *Lux*
		GaAs cell, Area $= 45\,\text{cm}^2$, output power $= 13.8\,\mu\,\text{W/cm}^2$ at 200 *Lux*

less [32]. Most of the custom design integrated circuits (ICs) (application circuit in Fig. 5.1) require a minimum of 1 V for operation. Also, many discrete ICs are available, which can be operated at low voltage [37–39], hence, can be powered by the energy harvesting system. Also, the power requirement of such ICs is more than the power harvested from the ambient. Therefore, it is not possible to directly power the application circuit by the energy harvester because of the voltage and power constraint. Therefore, an efficient boost converter is required, which boosts the output voltage of the energy harvester and stores the energy either on a battery or a storage capacitor [12, 14, 26, 31, 34, 40–48]. The stored energy on a storage element is given as input to a voltage regulator (DC-DC buck converter), which powers the application circuit. The time for storing the energy on a storage element is more than the time required by the application circuit to process/transmit the data gathered by the sensor [32]. For an efficient energy harvesting system, the power management block should be designed while considering the many aspects, which are discussed one by one.

5.1.1   Maximum Power Point Tracking Circuits

A simple electrical model (DC equivalent) of an energy harvester (or energy source) is shown in Fig. 5.2 [49]. The model contains voltage source Vs in series with resistance R_s and both in parallel with capacitance C_s [50]. The DC model of any energy harvester is used for designing the energy harvesting system. For example, an RF energy harvesting system needs rectifier at the input to convert the RF energy into DC energy. The rectifier used will be operating at the GHz frequency due to high-frequency RF signals. However, the entire energy harvesting system cannot be simulated at GHz frequency because it will take long time to complete the simulation. Therefore, the RF source with rectifier is replaced with a DC source (Fig. 5.2). This helps in simulating the energy harvesting system by decreasing the simulation time.

   In most cases, the power requirement of the application circuit is more than the power harvested from the ambient. Therefore, it is necessary to harvest the maximum power from the energy harvester and store it on a storage capacitor. According to maximum power transfer theorem for DC power source, the load resistance (R_L) should be equal to the source resistance (R_s) [51]. Another way of describing the maximum power transfer theorem is that load voltage (V_L) should be equal to half of the source voltage (V_s). The method used to achieve any of the above

Fig. 5.2   A simple model of an energy harvester followed by load

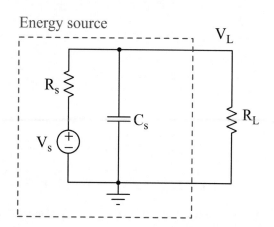

conditions is known as maximum power point tracking (MPPT) [12, 15, 41, 43, 52–55] implements the energy harvesting system without considering the MPPT. If the method used for MPPT is based on matching the source resistance with the load resistance, then it is known as resistor emulation method [1, 40, 45–48, 51, 56] use resistor emulation based MPPT. The resistor emulation method is based on open loop. Therefore, any variation in the parameters of the energy harvester or the power management circuit will have substantial effect on the maximum power point. Another method is based on maintaining the output voltage of the energy harvester equal to $V_s/2$ and is popularly known as fractional open-circuit-voltage (FOCV) based MPPT [14, 25, 27, 31, 35, 42, 48, 50, 57, 58]. FOCV based MPPT requires closed loop architecture. Hence, any change in the energy harvester and the power management circuit gets compensated by the closed loop. One more method popularly used for photovoltaic energy harvester is hill climbing approach (or perturb and observation) [3, 10, 26, 28, 30, 34, 59–63]. Instead of matching load resistance with the source resistance, or maintaining output of the energy harvester equal to $V_s/2$, the hill climbing approach tries to maximize the power in the load resistance.

5.1.1.1   Customized Boost and Buck Converters

The basic schematic of the proposed PMC for energy harvesting is shown in Fig. 5.3. The equivalent circuit of all types of energy harvesters is represented by a finite power source modeled by a voltage source V_s in series with resistance R_s and both in parallel with C_s. The start-up is an auxiliary energy harvester, which will build up the storage voltage V_{sto} up to 800 mV for the control circuit in the PMC to work.

As shown in Fig. 5.3, the proposed PMC circuit also constitutes voltage monitor circuits that monitor V_{sto}. One of them indicates when V_{sto} crosses 800 mV, then energy can be harvested from the main source (represented by V_s and R_s) and the

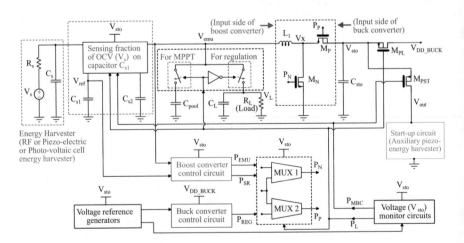

Fig. 5.3 Basic schematic of the proposed power management circuit (PMC)

boost converter will be enabled. The boost converter has one inductor L_1 and two switches M_N and M_P as shown in Fig. 5.3. For MPPT operation, the boost converter maintains its input voltage V_{emu} as a fraction (α) of the open circuit output voltage (V_s) of the energy harvester. The control circuit of the boost converter controls the switches M_N and M_P. The storage capacitor C_{sto} is further charged by the boost converter until Vsto reaches 1.8 V. At that moment, another voltage monitor circuit provides an indication to the buck converter for providing a constant voltage of 1 V across load resistance R_L. Most of the control circuit is designed using regular-Vt (or nominal-Vt) transistors. Therefore, the maximum storage voltage is chosen equal to 1.8 V. If storage voltage V_{sto} increases more than 1.8 V, a protection circuit can be provided at the V_{sto} node. The maximum storage voltage can be even less than or more than 1.8 V, and the design can be modified accordingly. The buck converter uses the same inductor and switches as used by the boost converter as shown in Fig. 5.3. Input to the buck converter is voltage V_{sto} and the supply for its control circuit is provided by V_{DD_BUCK}. Since the buck converter takes energy from the capacitor C_{sto}, voltage V_{sto} will decrease till it reaches 1 V, and the boost converter is again enabled. There is a need of multiplexer circuit that will select the corresponding control signals depending on if boost converter or buck converter is operational as shown in Fig. 5.3. This is done by multiplexers MUX 1 and MUX 2 with the select signal as output of voltage monitor circuit as shown in Fig. 5.3.

Figure 5.4 shows the voltage monitor circuit used for enabling either the boost converter or the buck converter depending on V_{sto}. Figure 5.4a shows a voltage reference generator that generates V_{ref1} ($\approx$240 mV) and V_{ref2} ($\approx$480 mV) voltage references [32, 64]. For reducing the current consumption of the reference generator, gate and source of low-Vt transistor M_{PLVT} are shorted so that it will operate in

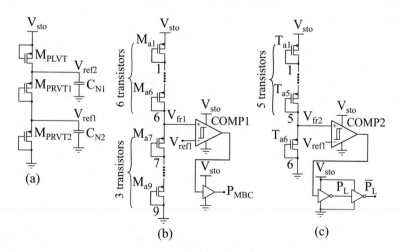

Fig. 5.4 Voltage monitor circuits. (**a**) Voltage reference generator. (**b**) Comparison of fraction of V_{sto} with V_{ref1} to generate P_{MBC}. (**c**) Comparison of fraction of V_{sto} with V_{ref2} to generate P_L and $\overline{P_L}$

sub-threshold region. The current flowing in M_{PLVT} is independent of supply V_{sto}. This current is passed through regular-Vt transistors M_{PRVT1} and M_{PRVT2}. The positive temperature coefficient of the difference in threshold voltage of M_{PRVT1} and M_{PLVT} is balanced by the negative temperature coefficient produced by the different sizing of transistors M_{PRVT1} and M_{PLVT} to produce temperature independent voltage references. The size of M_{PLVT} is 2.5 μm/50 μm and the size of both M_{PRVT1} and M_{PRVT2} is 6 μm/50 μm.

5.1.1.2   Start-Up Circuit and Voltage Monitoring

In the beginning of energy harvesting process, there is no initial charge on C_{sto} shown in Fig. 5.3 and hence all the signals in the power management circuit are LOW. One way of charging C_{sto} is using an auxiliary piezo-energy harvester. There is a PMOS switch (M_{PST}) connected between the auxiliary piezo-energy harvester output (V_{out}) and capacitor C_{sto}. M_{PST} is controlled by the signal P_{MBC}, which is LOW till V_{sto} reaches around 800 mV. Therefore, M_{PST} is ON during the start-up phase and hence C_{sto} is charged by the auxiliary piezo-energy harvester. The maximum current consumed by the entire control circuit during the start-up is only 14 nA, which does not load the auxiliary energy harvester much. Start-up circuits are also reported as a multistage charge-pump (or non-optimized inductor based boost converter), whose switching frequency is generated by a ring oscillator (coldstart) [12, 25, 42, 48, 50, 65–67]. When V_{sto} reaches around 800 mV, signal P_{MBC} goes HIGH and turns off switch M_{PST}.

Voltage monitor circuits, shown in Fig. 5.4b,c, are required to monitor V_{sto}. Figure 5.4b shows a hysteretic comparator [68] that compares a fraction (V_{fr1}) of V_{sto} with V_{ref1} and generates the control signal P_{MBC} that disables the start-up and enables the boost converter and of which the simulated result is shown in Fig. 5.5a. Comparator COMP1 has a hysteresis of around 200 mV. Figure 5.4c shows another voltage monitor circuit that makes P_L go HIGH as shown in Fig. 5.5b, when V_{sto} reaches 1.8 V to disable the boost converter and enable the buck converter. Comparator COMP2 has a hysteresis of around 800 mV. Dickson based charge-pump can also be used for the start-up, as shown in Fig. 5.6a. R_{CTRL} models the control circuit as a dc load and is approximately 100 MΩ. The 5- stage ring oscillator is powered by the output of the energy harvester to generate a clock for the charge-pump. CLK_{1SU} and $\overline{CLK_{1SU}}$ are alternatively connected to C_{CC1} down to C_{CC4}. CLK_{2SU} and $\overline{CLK_{2SU}}$ are alternatively connected to C_{CC5} down to C_{CC8}. Similarly, CLK_{3SU} and $\overline{CLK_{3SU}}$ are alternatively connected to C_{CC9} down to C_{CC12}. The value of all the coupling capacitances (C_{CC1}–C_{CC12}) is kept more than the parasitic capacitances (C_{PC1}–C_{PC12}) and (C_{int1}–C_{int6}), and is equal to 500 fF. The sizes of M_{CP1}–M_{CP12} are 6 μm/1 μm, and the PMOS and NMOS transistors in inverters (INV1–INV6) are (1 μm × 20)/180 nm and (1 μm × 10)/180 nm. M_{PST} is a thick oxide transistor, and its size is (100 μm × 10)/340 nm. The minimum value of V_s (OCV) is 350 mV for $R_s = 100\,k\Omega$, which is needed to generate 1 V across C_{sto}. After V_{sto} crosses 1 V, P_{MBC} goes high and the power from the energy harvester is harvested using the MPPT.

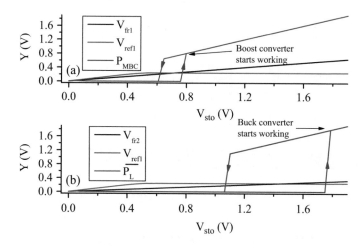

Fig. 5.5 Simulation results of the control operation by the voltage monitor circuits of Fig. 5.4. (**a**) Signal P_{MBC} goes HIGH when V_{sto} reaches 800 mV. (**b**) Signal $\overline{P_L}$ goes HIGH when V_{sto} reaches 1.8 V

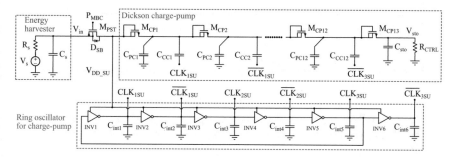

Fig. 5.6 Start-up circuit. (**a**) Energy harvester followed by 12-stage Dickson based charge-pump. R_{CTRL} models the control circuit as a dc load. (**b**) Ring oscillator for generating clock for the charge-pump

5.1.2   Optimal Design of Boost Converter for MPPT

DC-DC boost converter is used at the output of the rectifier in case of RF and piezo-energy harvesters, and at the output of the solar cell in case of solar energy harvester as shown in Fig. 5.7. The input voltage (V_{emu}) of the boost converter is maintained at a fraction (Ω) of the open circuit voltage (V_s) of the respective energy harvester [1] for MPPT. P_{avl} ($V_s^2/4 \times R_s$) for $\alpha = 0.5$ is the maximum power that can be harvested from the energy harvester and provided in the boost converter. $C_{pool}(>> C_s)$ is a large capacitor connected between the harvester and the boost converter to act as the first storage element. The boost converter optimally transfers the energy from C_{pool} to C_{sto} and also boosts its input voltage V_{emu}. C_{pool} should be large enough so that the ripple in voltage V_{emu} is kept less than 10% of V_{emu}

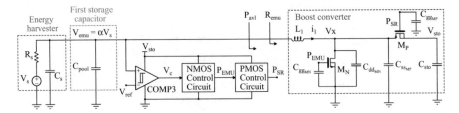

Fig. 5.7 The energy harvester followed by a first storage capacitor C_{pool} and boost converter. $C_{gg_{MN}}, C_{dd_{MN}}, C_{ss_{MP}}, C_{gg_{MP}}$ are parasitic capacitances associated with the boost converter. P_{EMU} and P_{SR} signals are generated by the control circuit

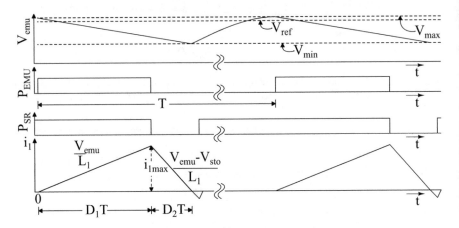

Fig. 5.8 V_{emu} is the input voltage of the boost converter. P_{EMU} and P_{SR} signals controlling the switches of the boost converter and the inductor current flowing in the boost converter

during the MPPT. The boost converter switches are controlled by signals P_{EMU} and P_{SR}. P_{EMU} helps to achieve the MPPT at the input of the boost converter. P_{SR} helps in achieving the synchronous rectification in the boost converter so that there is no reverse flow of current from C_{sto} to the energy harvester. The control circuits for the generation of P_{EMU} and P_{SR} signals are powered by V_{sto}. For implementing the MPPT and synchronous rectification technique, analysis of the boost converter is required. First the input resistance of the boost converter is calculated for discontinuous conduction mode. After that, the power loss in the boost converter is minimized, which gives the time period and duty cycle of the signal P_{EMU} and their dependence on the boost ratio ($B_R = V_{sto}/V_{emu}$). Finally design of the circuit realizing P_{EMU} and P_{SR} is presented.

Figure 5.8 shows the timing diagram of the P_{EMU} and P_{SR} signals controlling the switches and the inductor current (i_1) of the boost converter. D_1 and T are the duty cycle and time period of P_{EMU} signal. The input resistance R_{emu} of the boost converter is given by (5.1).

$$R_{emu} = \frac{V_{emu}}{\overline{(i_1)}} = \frac{2L_1}{(D_1)^2 T}\left(1 - \frac{1}{B_R}\right) \tag{5.1}$$

Energy dissipation in the boost converter is in two forms, ohmic loss and switching loss.

5.1.2.1 Ohmic Loss Calculation

In Fig. 5.8, during $0 < t < D_1 T$, M_N is ON, and the energy is stored in the inductor L_1. Inductor current i_1 will dissipate power P_{R1} in the DC series resistance (R_{ind}) of L_1, ON resistance (R_N) of M_N and equivalent series resistance (R_C) of C_{pool} as given by (5.2) and (5.3).

$$i_1 = \frac{V_{emu}}{L_1}t \tag{5.2}$$

$$P_{R1} = \frac{dE_{R1}}{dt} = i_1^2 R_1 \tag{5.3}$$

In (5.3), $R_1 = R_{ind} + R_N + R_C$. Total loss ($E_{R1}$) during this time is calculated by integrating (5.3) and is given in (5.4). Duty cycle D_1 from (5.1) is substituted in (5.4) and the final expression for E_{R1} during inductor energizing is given by (5.5).

$$E_{R1} = \frac{1}{3}\left(\frac{V_{emu}}{L_1}\right)^2 R_1 (D_1 T)^3 \tag{5.4}$$

$$E_{R1} = \frac{1}{3}\left(\frac{V_{emu}}{L_1}\right)^2 T^{\frac{3}{2}}\left(\frac{2L_1}{R_{emu}}\left(1 - \frac{1}{B_R}\right)\right)^{\frac{3}{2}} R_1 \tag{5.5}$$

During $D_1 T < t < (D_1 + D_2)T$, M_N is OFF and M_P is ON and the stored energy in L_1 is transferred to C_{sto}. Inductor current and power dissipated during this period is given by (5.6) and (5.7). V_{sto} is almost constant because of large value of C_{sto}.

$$i_1 = i_{1max} - \left(\frac{v_{emu} - V_{sto}}{L_1}\right)(D_1 T - t) \tag{5.6}$$

$$P_{R2} = \frac{dE_{R2}}{dt} = i_1^2 R_2 \tag{5.7}$$

In (5.7), $R_2 = R_{ind} + R_P + R_C$, where R_P is ON resistance of M_P. The final expression for E_{R2} is given by (5.8). The average power loss $P_R = (E_{R1} + E_{R2})/T$, is given by (5.9).

$$E_{R2} = \frac{1}{3}\left(\frac{V_{emu}}{L_1}\right)^2 T^{\frac{3}{2}}\left(\frac{2L_1}{R_{emu}}\left(1 - \frac{1}{B_R}\right)\right)^{\frac{3}{2}}\left(\frac{R_2}{B_R - 1}\right) \tag{5.8}$$

$$P_R = \frac{1}{3}\left(\frac{V_{emu}}{L_1}\right)^2 \sqrt{T}\left(\frac{2L_1}{R_{emu}}\left(1 - \frac{1}{B_R}\right)\right)^{\frac{3}{2}}\left(R_1 + \frac{R_2}{B_R - 1}\right) \tag{5.9}$$

5.1.2.2 Switching Loss

The total effective capacitance (C_{eff}) will be the sum of all the parasitic capacitances $(C_{gg\_MN}, C_{dd\_MN}, C_{gg\_MP}, C_{ss\_MP})$ shown in Fig. 5.7. The total switching loss is given by (5.10).

$$P_{SW} = C_{eff}V_{sto}^2\frac{1}{T} \tag{5.10}$$

5.1.2.3 Minimizing Total Loss

Total power loss (P_{loss}) in the boost converter is given by the addition of (5.9) and (5.10). With the increase in the switching time period T, ohmic loss increases while switching loss decreases. The expression of T at which the total loss is minimum is found by differentiating P_{loss} with respect to T and given by (5.11). Substituting T_{opt} in (4.1), $(D_1T)_{opt}$ can be found, which is given by (5.12). Also the time duration for which M_P is ON is given by $(D_2T = (D_1T)/(B_R - 1))$ and finally is given by (5.13).

$$T_{opt} = \left(\frac{6C_{eff}B_R^2L_1^2}{\left(\frac{2L_1}{R_{emu}}\left(\frac{B_R-1}{B_R}\right)\right)^{\frac{3}{2}}\left(R_1 + \frac{R_2}{B_R-1}\right)}\right)^{\frac{2}{3}} \tag{5.11}$$

$$(D_1T)_{opt} = \left(\frac{6C_{eff}B_R^2L_1^2}{R_1 + \frac{R_2}{B_R-1}}\right)^{\frac{1}{3}} \tag{5.12}$$

$$(D_2T)_{opt} = \left(\frac{6C_{eff}L_1^2}{R_1 + \frac{R_2}{B_R-1}}\right)^{\frac{1}{3}}\frac{B_R^{\frac{2}{3}}}{B_R - 1} \tag{5.13}$$

5.1.2.4 MPPT Implementation by the Boost Converter

Maximum power needs to be harvested from the energy harvester by the boost converter controlled by signals P_{EMU} and P_{SR} as shown in Fig. 5.7. Time period T_{opt} of P_{EMU} signal helps in minimizing the loss in the boost converter. $(D_1 T)_{opt}$ is the ON time for M_N and helps in achieving the MPPT between the energy harvester and the boost converter. $(D_2 T)_{opt}$ is the ON time for $M_P 4$ and helps to stop the reverse flow of energy from C_{sto} to C_{pool}. Input voltage (V_{emu}) of the boost converter is compared with the reference voltage $(V_{ref} = \alpha \times V_s)$ using hysteretic comparator COMP3. The value of α lies between 0.45 to 0.55 for the RF and piezo-energy harvester and 0.75 to 0.85 for the solar energy harvester. As soon as V_{emu} is more than V_{ref}, M_N is turned-on with the help of NMOS control circuit. The time duration for which M_N is ON is given by 5.12. After the NMOS is turned-off, M_P is turned-on and its ON time is given by (5.13).

Clearly, $T_{opt} \propto (B_R)^{4/3}$, $(D_1 T)_{opt} \propto (B_R)^{2/3}$ and $(D_2 T)_{opt} \propto (B_R)^{2/3}/(B_R - 1)$. It is not easy to generate these signals and hence, for simplicity some approximations are considered, which are given as $T_{opt} \propto (B_R)^2$, $(D_1 T)_{opt} \propto B_R$ and $(D_2 T)_{opt} \propto B_R/(B_R - 1)$. Based on these approximations, the approximate relations for T_{opt}, $(D_1 T)_{opt}$, and $(D_2 T)_{opt}$ are given by (5.14), (5.15), and (5.16). The total minimal power loss (P_{loss_min}) in the boost converter, which is found by substituting (5.14) in the expression of P_{loss} is given by (5.17). With these approximations, for $R_{emu} = 65 \, k\Omega$, the normalized power loss with respect to the maximum available power $(P_{norm} = P_{loss\_min}/(V_{emu}^2/R_{emu}))$ in the boost converter will increase from 8.31% to 12.54% for Pavl $= 153 \, nW$ and from 4.42% to 5.07% for Pavl $= 1.38 \, \mu W$, without losing the concept of MPPT.

$$
T_{opt} = \left(\frac{6 C_{eff} L_1^2}{\left(\frac{2L_1}{R_{emu}} \left(\frac{B_R - 1}{B_R} \right) \right)^{\frac{3}{2}} \left(R_1 + \frac{R_2}{B_R - 1} \right)} \right)^{\frac{2}{3}} B_R^2 \tag{5.14}
$$

$$
(D_1 T)_{opt} = \left(\frac{6 C_{eff} L_1^2}{R_1 + \frac{R_2}{B_R - 1}} \right)^{\frac{1}{3}} B_R \propto \frac{V_{sto}}{V_{emu}} \tag{5.15}
$$

$$
(D_2 T)_{opt} = \left(\frac{6 C_{eff} L_1^2}{R_1 + \frac{R_2}{B_R - 1}} \right)^{\frac{1}{3}} \frac{B_R}{B_R - 1} \propto \frac{V_{sto}}{V_{sto} - V_{emu}} \tag{5.16}
$$

$$
P_{loss\_min} = \left(1.21 + \frac{0.605}{B_R} \right) \left(\frac{V_{emu}^2}{R_{emu}} \right) \left(\frac{C_{eff}}{L_1} \right)^{\frac{1}{3}} \times
$$
$$
\left(R_1 + \frac{R_2}{B_R - 1} \right)^{\frac{2}{3}} (B_R - 1) \tag{5.17}
$$

For the commercial inductor $L_1 = 820\,\mu H$, its DC series resistance is $R_{ind} = 2.7\,\Omega$ [69]. P_{loss_min} depends on the resistance and parasitic capacitance of the switches, which in turn depends on their size. P_{norm} is plotted for different sizes of M_N and M_P as shown in Fig. 5.9. For plotting Fig. 5.9, $V_s = 2 \times V_{emu} = 200\,mV$, $R_s = R_{emu} = 65\,k\Omega$, $V_{sto} = 1\,V$, and $R_C = 1\,W$ are chosen. R_C includes the equivalent series resistance of $C_{pool} = 10\,\mu F$ [70] and switch resistance, which are in series with C_{pool}. The reason for introducing switches in series with C_{pool} will be discussed in Fig. 5.14. The sizes at which P_{norm} is minimum (12.54%) are 2.5 mm for M_N and 1 mm for M_P in 180 nm mixed-mode CMOS process. Therefore, $R_N = 2.31\Omega$, $R_P = 13.8\Omega$, and $C_{eff} = 17.06\,pF$ at typical corner.

5.1.2.5 Sensing Vref Voltage

V_{ref} is a fraction of the open circuit voltage of the energy harvester. The circuit used for sensing V_{ref} with the help of control signals S_1 and S_2 is shown in Fig. 5.10a. Initially the boost converter is disconnected from the energy harvester and its open circuit voltage (V_s) gets accumulated over C_{s1} by turning-on SW_1 using S_1 for

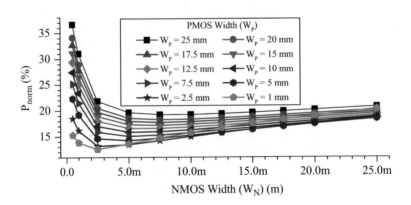

Fig. 5.9 Optimal power loss normalized with respect to available power and plotted for different sizes of NMOS and PMOS switches

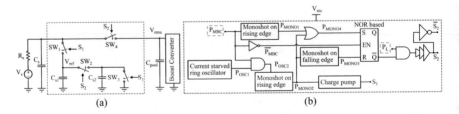

Fig. 5.10 (a) The circuit designed using switches and control signals S_1 and S_2 to sense fraction of open circuit voltage V_s. (b) The control circuit to generate signals S_1 and S_2. P_L is HIGH till V_{sto} reaches 1.8 V. P_{MBC} and P_L are provided by the voltage monitor circuits

the duration of $T_s = 10$ ms. T_s should be around five times of the time constant introduced by resistor R_s and capacitor C_{s1}. During this time, the voltage across C_{s2} gets discharged by SW_3. After 10 ms, SW_1 and SW_3 are turned-off and SW_2 and SW_4 are turned-on by S_2. Therefore, C_{s1} and C_{s2} come in parallel and the initial energy stored on C_{s1} will be distributed across C_{s1} and C_{s2} depending on their values. S_2 will remain HIGH for 4.5 s and provides the necessary V_{ref} across C_{s1} for operation of the boost converter. Maintaining V_{emu} equal to V_{ref} helps in implementing the MPPT at the input of the boost converter. The control circuit used for generating S_1 and S_2 and its timing diagram is shown in Fig. 5.10b.

5.1.2.6 $(D_1T)_{opt}$ Implementation

The circuit used for generating $(D_1T)_{opt}$ is shown in Fig. 5.11. Operation of the complete circuit consists of two phases, start-up and MPPT. During start-up, V_{sto} is charged by the start-up circuit and the control circuit shown in Fig. 5.11a–c is OFF. During this phase, outputs of the SR latches are LOW because P_{MBC} is

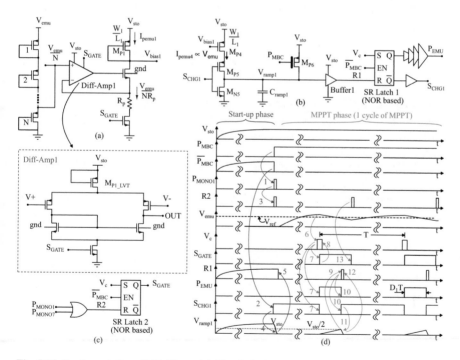

Fig. 5.11 Implementation of $(D_1T)_{opt}$. (**a**) Circuit designed to produce current proportional to V_{emu}. (**b**) Producing ON time of M_N proportional to V_{sto}/V_{emu}. (**c**) Circuit designed to generate a signal that will allow current flow in Fig. 5.11a and 5.12a only for $((D_1T)_{opt} + (D_2T)_{opt})$ time. (**d**) Timing diagram associated with the complete circuit. V_{sto} is assumed constant for one cycle of MPPT

HIGH. At the beginning of the MPPT phase, P_{MBC} goes HIGH, which generates a monoshot pulse PMONO1 as shown in Fig. 5.10b, which resets SR latch 2 in Fig. 5.11c. Hence, no current is consumed by the circuit in Fig. 5.11a because S_{GATE} is LOW. C_{pool} is charged by the energy harvester and as soon as V_{emu} reaches more than V_{ref}, signal V_c goes HIGH as shown in Fig. 5.7. It indicates that M_N should be turned-on for $(D_1T)_{opt}$ time. The rising edge of V_c sets the SR Latch 1 and SR Latch 2 in Fig. 5.11b,c and signals P_{EMU} and S_{GATE} go HIGH and S_{CHG1} goes LOW. P_{EMU} turns-on M_N and S_{GATE} allows the circuit of Fig. 5.11a to work. A fraction of the V_{emu} is sensed and given as input to a differential amplifier, shown in Fig. 5.11a, which generates a current proportional to V_{emu} ($I_{pemu1} = V_{emu}/(N \times R_P)$) passing through M_{P1} transistor. $N = 6$ is chosen here to reduce the loading on V_{emu} and $R_p = 333k\Omega$ is chosen so that low current will flow through M_{P1} transistor. Current from M_{P1} is mirrored into M_{P4} ($I_{pemu4} = I_{pemu1}$) in Fig. 5.11b. Since S_{CHG1} is LOW, M_{P5} is ON therefore, I_{pemu4} charges C_{ramp1} gradually ($V_{ramp1} = (Ipemu4 \times t)/C_{ramp1}$). As soon as the voltage across C_{ramp1} reaches $V_{sto}/2$, output of Buffer 1 (R1) goes HIGH and resets the SR Latch 1. Therefore, P_{EMU} goes LOW and MN turns off. Also S_{CHG1} goes HIGH and C_{ramp1} discharges through M_{N5} from $V_{sto}/2$ to 0. The time for which P_{EMU} is HIGH is also the time for which C_{ramp1} is charged from 0 to $V_{sto}/2$ and is given by (5.18).

$$t_{0 \rightarrow V_{sto/2}} = \left(\frac{N}{2}\right) C_{ramp1} R_p \left(\frac{V_{sto}}{V_{emu}}\right) \tag{5.18}$$

Clearly, this time is proportional to the ratio V_{sto}/V_{emu} as given in (5.15). From (5.15), for $V_{emu} = 100\,mV$ and $V_{sto} = 1\,V$, $(D_1T)_{opt}$ is $20.52\,\mu s$. Therefore $t_{0 \rightarrow V_{sto/2}}$ should be equal to $20.52\,\mu s$ in (5.18). Substituting all the known parameters in (5.18), C_{ramp1} is found to be 2.05 pF. As soon as P_{EMU} goes LOW, stored energy in the inductor should be transferred to C_{sto} through M_p (Fig. 5.7). Therefore, P_{SR} needs to go LOW as soon as P_{EMU} goes LOW [49]. The circuit to generate P_{SR} is shown in Fig. 5.12. The time for which P_{SR}

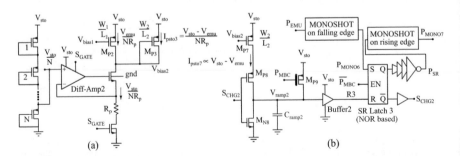

Fig. 5.12 Implementation of $(D_2T)_{opt}$. (**a**) Circuit designed to produce current proportional to $V_{sto} - V_{emu}$. (**b**) Producing ON time of M_P proportional to $V_{sto}/(V_{sto} - V_{emu})$

is LOW is given by (5.16). At the falling edge of P_{EMU}, monoshot signal P_{MONO6} is generated, which sets the SR Latch 3 and therefore P_{SR} and S_{CHG2} go LOW. A fraction of V_{sto} is sensed and is given as input to the differential amplifier 2 in Fig. 5.12a, which generates a current equal to $I_{psto3} = (V_{sto} - V_{emu})/(N \times R_P)$ and flows through M_{P3}. Current from M_{P3} is mirrored into the M_{P7} in Fig. 5.12b. Since S_{CHG2} is LOW, M_{P8} is ON and hence I_{psto7} gradually charges C_{ramp2} ($V_{ramp2} = (I_{psto6} \times t)/C_{ramp2}$). As soon as V_{ramp2} reaches $V_{sto2}/2$, output of Buffer 2 (R3) goes HIGH and resets the SR Latch 3. Therefore, P_{SR} goes HIGH and M_P turns off. Also S_{CHG2} goes HIGH and C_{ramp2} discharges through M_{N8} from $V_{sto}/2$ to 0. The time for which P_{SR} is LOW is also the time for which C_{ramp2} is charged from 0 to $V_{sto}/2$ and is given by 5.19.

$$t'_{0 \to V_{sto/2}} = \left(\frac{N}{2}\right) C_{ramp2} R_p \left(\frac{V_{sto}}{V_{sto} - V_{emu}}\right) \tag{5.19}$$

5.1.2.7 Synchronous Rectification Control for the Boost Converter

Clearly this time is proportional to $V_{sto}/(V_{sto} - V_{emu})$ as given in (5.16). From (5.16), for $V_{emu} = 100\,\text{mV}$ and $V_{sto} = 1\,\text{V}$, $(D_2T)_{opt}$ is 2.28 µs. Therefore $t'_{0 \to V_{sto/2}}$ should also be equal to 2.28 µs in (5.19). Substituting all the known parameters in (5.19), Cramp2 is found to be 2.05 pF.

5.1.2.8 Calculation of the First Storage Capacitor, C_{pool}

During the time when both MN and MP are turned off, C_{pool} is charged from V_{min} to V_{max} in time $T_{opt}(T_{opt} >> (D_1T)_{opt} + (D_2T)_{opt})$ and voltage V_{emu} is given by (5.20). For MPPT, average voltage of V_{emu} is a fraction (α) of the open circuit voltage (V_s) of the energy harvester. C_{pool} is large enough so that the ripple in V_{emu} is kept less than 10% of V_{emu}. Therefore, V_{min} is equal to $0.9 \times \alpha \times V_s$. After substituting all the known parameters in (5.20), C_{pool} is given by (5.21).

$$V_{emu}(t) = V_{min} + (V_s - V_{min})\left(1 - exp\left(\frac{-t}{C_{pool} \times R_s}\right)\right) \tag{5.20}$$

$$C_{pool} = \frac{T_{opt}}{R_s \times ln\left(\frac{1-0.9\alpha}{1-\alpha}\right)} \tag{5.21}$$

For the RF and piezo-energy harvester, $R_{emu} \approx R_s$, $C_{s1} \approx C_{s2}$ and $\alpha \approx 0.5$. Therefore, T_{opt}/R_s is calculated from Eq. (5.14) and substituted in (5.21), which gives (5.22). For $V_{emu} = 50\,\text{mV}$ and $V_{sto} = 1\,\text{V}$, $C_{pool} \approx 13\,\mu\text{F}$.

$$C_{pool} = \frac{1}{0.0953} \left(\frac{6C_{eff}L_1^2}{\left(2L_1\left(\frac{B_R-1}{B_R}\right)\right)^{\frac{3}{2}}\left(R_1 + \frac{R_2}{B_R-1}\right)} \right)^{\frac{2}{3}} B_R^2 \qquad (5.22)$$

For the solar energy harvester, $R_{emu} \approx 4 \times R_s$, $C_{s1} \approx 4 \times C_{s2}$ and $\alpha \approx 0.8$; however, source resistance R_s is not constant and depends on the irradiance of light. For $P_{irr} = 0.9\,\text{mW/cm}^2$, maximum value of P_{avl} is $342\,\mu\text{W}$ at $V_{emu} = 355\,\text{mV}$. This gives $R_{emu} = 370\,\Omega$ and $R_s = 92\,\Omega$. For $V_{sto} = 1\,\text{V}$, T_{opt} from (5.14) is $7.4\,\mu\text{s}$ and C_{pool} from (5.21) is $240\,\text{nF}$.

5.1.3   Design of the Buck Converter as Output Voltage Regulator

As soon as V_{sto} reaches 1.8 V, P_L goes LOW. This turns-on the PMOS switch M_{PL} to provide the supply voltage $V_{DD\_BUCK}(= V_{sto})$ to the control circuit of the buck converter as shown in Fig. 5.13. This arrangement prevents the power loss in the control circuit of the buck converter during MPPT. The buck converter now provides a regulated voltage across a load (R_L) connected at its output as shown in Fig. 5.13a. The buck converter is designed to operate at the boundary of continuous conduction mode and discontinuous conduction mode. Here, inductor and switches are same as those used in the design of the boost converter. Now the only parameters that need to be found are the value of capacitor C_L and the delay of comparator COMP4 for a particular load resistance R_L. For the design purpose nominal load (R_L) and voltage (V_L) are assumed to be $1\,\text{k}\Omega$ and 1 V, respectively. The average value of the inductor current (I_{2avg}) is equal to the load current as shown in Fig. 5.13b. Also, average value of inductor current in terms of maximum inductor current (I_{2max}) is

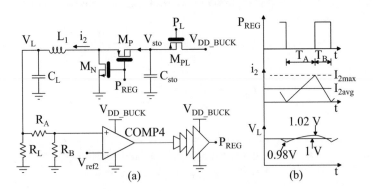

Fig. 5.13 (a) Buck converter used as regulator. (b) Buck converter waveforms in steady state

given by $0.5 \times I_{2max}$. Hence, I_{2max} is equal to 2 mA. When M_P is ON, inductor current will increase from 0 to I_{2max} in time T_A and voltage across the inductor remains at 0.8 V given by,

$$V_{L2} = L_1 \frac{di_2}{dt} = L_1 \frac{I_{2max}}{T_A} \tag{5.23}$$

After substitution, T_A comes out to be 2.05 μs. Therefore, delay of the comparator is 2.05 μs. When M_N is ON, inductor current will decrease from I_{2max} to 0 in time T_B and the voltage across the inductor is -1 V. After substituting the values in (5.23), T_B comes out to be 1.64 μs. Because of finite delay of the comparator there is ripple across the load R_L. To have a moderate ripple in the range of 0.98 V to 1.02 V, there is a need of finite capacitance C_L across R_L. Energy consumed by the load during time T_B is given by (5.24).

$$E_L = \frac{V_L^2}{R_L} \times T_B \tag{5.24}$$

This energy is supplied by the capacitor C_L and the voltage across C_L will change from 1.02 V to 0.98 V. Therefore, change in energy in the capacitor is given by (5.25). Equating (5.24) and (5.25), value of capacitor C_L is found to be 40 nF.

$$\delta E_C = \frac{1}{2} \times C_L \times (1.02^2 - 0.98^2) \tag{5.25}$$

The buck converter is designed for 1 mA of nominal load current (I_L) and 1 V of nominal load voltage (V_L) while its input voltage (V_{sto}) changes from 1.8 V to 1.1 V. The line regulation estimated for the designed buck converter is 31 mV/V. For the load current variations from 0.5 mA to 10 mA and the input voltage equal to 1.8 V, the estimated load regulation is 1.05 mV/mA.

5.1.4 Inductor Sharing Between MPPT and Voltage Regulator

During MPPT, inductor (L_1) and switches (M_N and M_P) are used by the boost converter and during voltage regulation by the buck converter. Hence, there is need for a control circuit that will share them between the two converters and shown in Fig. 5.14. As already explained in Fig. 5.4, the boost converter will work when V_{sto} is more than 800 mV but less than 1.8 V, which means P_{MBC} and P_L signals are HIGH. As soon as V_{sto} reaches 1.8 V, P_L goes LOW and the buck converter starts working, which will bring the voltage V_{sto} down to 1 V and P_L will become HIGH again.

The boost converter needs two signals P_{EMU} and P_{SR} for controlling the switches M_N and M_P while the buck converter needs only one signal P_{REG}.

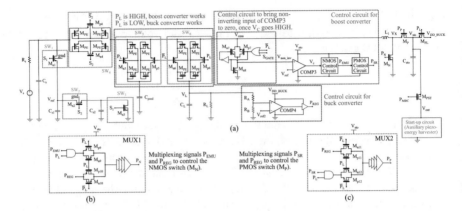

Fig. 5.14 Top level schematic showing the inductor (L_1) and switches (M_N and M_P) shared between the boost converter and the buck converter. The body of transistors M_{n4}, M_{n5}, and M_{n6} in the switches SW_4, SW_5, and SW_6, respectively, is connected to the lowest of potential that appears on source and drain of the respective switch. The body of transistors M_{p4}, M_{p5}, and M_{p6} in the switches SW_4, SW_5, and SW_6 respectively is connected to the highest of potential that appears on source and drain of the respective switch

Table 5.2 Switches size used for the circuit shown in Fig. 5.14

Switch	Type of transistors	Size (W/L) (m/m)
M_{n7}, M_{n9}, M_{n10}, M_{n11}, M_{n12}	Triple well high V_t	240 n/340 n
M_{PE}, M_{p7}, M_{p9}, M_{p10}, M_{p11} M_{p12}, M_{n1}, M_{n3}, M_{n8},	High V_t	240 n/340 n
M_{n2}, M_{PST}	High V_t	5 μ/340 n
M_{n4}, M_{n6}	Triple well high V_t	1 m/340 n
M_{n5}	Triple well high V_t	5 m/340 n
M_{p4}, M_{p6}, M_P, M_{PL},	High V_t	1 m/340 n
M_{p5}	High V_t	5 m/340 n
M_N	High V_t	2.5 m/340 n

Therefore, two multiplexers are needed whose outputs control the gates of two switches as shown in Fig. 5.14b,c. P_{EMU} is multiplexed with P_{REG} and P_{SR} is multiplexed with P_{REG} using multiplexers as shown in Fig. 5.14b,c; P_L and $\overline{P_L}$ are the control signals to these multiplexers. When P_L is HIGH, $P_N = P_{EMU}$ and $P_P = P_{SR}$ and the boost converter will be activated. When P_L is LOW, $P_N = P_P = P_{REG}$ and the buck converter will be activated.

During the operation of the boost converter, there is a need to bring the non-inverting input of comparator COMP3 to zero after V_c goes high, due to the hysteresis. When V_c goes HIGH, then S_{GATE} will be HIGH as shown in Fig. 5.11c and transistor M_{n8} will discharge the node V_{non_inv}, bringing back V_c to LOW. The sizes of the different switches used in Fig. 5.14 are given in Table 5.2. The value of different external components used in Fig. 5.14 is given in Table 5.3.

Table 5.3 External components values used in Fig. 5.14

Components	C_{pool}	C_{sto}	L_1	C_L	C_{s1}	C_{s2}
Value	$\approx$10 μF, depends on the energy harvester	1 μF for simulation	820 μH	$\approx$40 nF	50 nF	50 nF

References

1. S. Shahab, A. Erturk, Contactless ultrasonic energy transfer for wireless systems: acoustic-piezoelectric structure interaction modeling and performance enhancement. Smart Mater. Struct. **23**(12), 125032 (2014)
2. F. Touati, A. Galli, D. Crescini, P. Crescini, A.B. Mnaouer, Feasibility of air quality monitoring systems based on environmental energy harvesting, in *2015 IEEE International Instrumentation and Measurement Technology Conference (I2MTC) Proceedings* (2015), pp. 266–271
3. J. Lu, L. Zhang, S. Matsumoto, H. Hiroshima, K. Serizawa, M. Hayase, T. Gotoh, Miniaturization and packaging of implantable wireless sensor nodes for animals monitoring, in *2016 IEEE SENSORS* (2016), pp. 1–3
4. J. Lu, H. Okada, T. Itoh, R. Maeda, T. Harada, Towards the world smallest wireless sensor nodes with low power consumption for 'green' sensor networks, in *SENSORS, 2013 IEEE* (2013), pp. 1–4
5. L. Wang, G. Zhang, H. Liu, T. Hang, The network nodes design of gas wireless sensor monitor, in *2011 Second International Conference on Mechanic Automation and Control Engineering* (2011), pp. 1778–1781
6. D. Kissinger, A. Schwarzmeier, F. Grimminger, J. Mena-Carrillo, W. Weber, G. Hofer, G. Fischer, R. Weigel, Wireless integrated sensor nodes for indoor monitoring and localization, in *2015 IEEE Topical Conference on Wireless Sensors and Sensor Networks (WiSNet)* (2015), pp. 7–10
7. D.I. Sacaleanu, L.A. Perisoara, E. Spataru, R. Stoian, Low-cost wireless sensor node with application in sports, in *2017 IEEE 23rd International Symposium for Design and Technology in Electronic Packaging (SIITME)* (2017), pp. 395–398
8. J. Lu, H. Okada, T. Itoh, R. Maeda, T. Harada, Assembly of super compact wireless sensor nodes for environmental monitoring applications, in *2013 Symposium on Design, Test, Integration and Packaging of MEMS/MOEMS (DTIP)* (2013), pp. 1–4
9. R.C. Jisha, M.V. Ramesh, G.S. Lekshmi, Intruder tracking using wireless sensor network, in *2010 IEEE International Conference on Computational Intelligence and Computing Research* (2010), pp. 1–5
10. R.J.M. Vullers, R.v. Schaijk, H.J. Visser, J. Penders, C.V. Hoof, Energy harvesting for autonomous wireless sensor networks. IEEE Solid-State Circuits Mag. **2**(2), 29–38 (2010)
11. G. Papotto, F. Carrara, A. Finocchiaro, G. Palmisano, A 90-nm CMOS 5-mbps crystalless RF-powered transceiver for wireless sensor network nodes. IEEE J. Solid State Circuits **49**(2), 335–346 (2014)
12. Y. Huang, T. Tzeng, T. Lin, C. Huang, P. Yen, P. Kuo, C. Lin, S. Lu, A self-powered CMOS reconfigurable multi-sensor SoC for biomedical applications. IEEE J. Solid State Circuits **49**(4), 851–866 (2014)
13. L. Xia, J. Cheng, N.E. Glover, P. Chiang, 0.56 v, −20 dbm RF-powered, multi-node wireless body area network system-on-a-chip with harvesting-efficiency tracking loop. IEEE J. Solid State Circuits **49**(6), 1345–1355 (2014)
14. A. Klinefelter, N.E. Roberts, Y. Shakhsheer, P. Gonzalez, A. Shrivastava, A. Roy, K. Craig, M. Faisal, J. Boley, S. Oh, Y. Zhang, D. Akella, D.D. Wentzloff, B.H. Calhoun, A 6.45 μw self-powered IoT SoC with integrated energy-harvesting power management and ULP asymmetric radios, in *2015 IEEE International Solid-State Circuits Conference – (ISSCC) Digest of Technical Papers* (2015), pp. 1–3. 124

15. M. Tabesh, N. Dolatsha, A. ____abian, A.M. Niknejad, A power-harvesting pad-less millimeter-sized radio. IEEE J. Solid S⌐ Circuits **50**(4), 962–977 (2015)

16. G. Papotto, F. Carrara, A. ⊟cchiaro, G. Palmisano, A 90nm CMOS 5mb/s crystalless RF transceiver for RF-powerec⌐/SN nodes, in *2012 IEEE International Solid-State Circuits Conference* (2012), pp. 452—4

17. W. Wang, N. Wang, E. Jaf⊸M. Hayes, B. O'Flynn, C. O'Mathuna, Autonomous wireless sensor network based build▮ energy and environment monitoring system design, in *2010 The 2nd Conference on En⸺nmental Science and Information Application Technology*, vol. 3 (2010), pp. 367–372

18. H. Song, P. Kumar, D. M ya, M. Kang, W.T. Reynolds, D. Jeong, C. Kang, S. Priya, Ultra-low resonant piezoel⊏ic mems energy harvester with high power density. J. Microelectromech. Syst. **26**(6), 12═-1234 (2017)

19. J.A. Hagerty, F.B. Helmbre$_{rb}$, W.H. McCalpin, R. Zane, Z.B. Popovic, Recycling ambient microwave energy with br$_{tate}$-band rectenna arrays. IEEE Trans. Microwave Theory Tech. **52**(3), 1014–1024 (2004) r_{in}

20. H.-B. Fang, J.-Q. Liu, Z.-Y.$_1$ ·, L. Dong, L. Wang, D. Chen, B.-C. Cai, Y. Liu, Fabrication and performance of MEMS-bas$_4$ piezoelectric power generator for vibration energy harvesting. Microelectron. J. **37**(11), 12$_{er}$ -1284 (2006). http://www.sciencedirect.com/science/article/pii/S0026269206001911 ·$_i$

21. M. Wahbah, M. Alhawari, ·$_i$ Mohammad, H. Saleh, M. Ismail, Characterization of human body-based thermal and vit ion energy harvesting for wearable devices. IEEE J. Emerging Sel. Top. Circuits Syst. **4**, 3⁼ 363 (2014)

22. V. Leonov, Thermoelectric ⸺rgy harvesting of human body heat for wearable sensors. IEEE Sensors J. **13**(6), 2284–229⸳ 013)

23. B.M. Kayes, L. Zhang, R. ⸳ st, I. Ding, G.S. Higashi, Flexible thin-film tandem solar cells with > 30% efficiency. J. Pl⸳ voltaics **4**(2), 729–733 (2014)

24. I. Mathews, P.J. King, F. St ord, R. Frizzell, Performance of iii-v solar cells as indoor light energy harvesters. IEEE J. F⸳tovoltaics **6**(1), 230–235 (2016)

25. P. Hsieh, C. Chou, T. Chiang n RF energy harvester with 44.1 power of -12 dbm. IEEE Trans. Circuits Syst. I Regl. Pap. **6**$_{0}$), 1528–1537 (2015)

26. Y. Qiu, C. Van Liempd, B.O. et Veld, P.G. Blanken, C. Van Hoof, 5µw-to-10mw input power range inductive boost conv er for indoor photovoltaic energy harvesting with integrated maximum power point trac$_{ra}$ g algorithm, in *2011 IEEE International Solid-State Circuits Conference* (2011), pp. 118⸺4 0

27. H. Chen, Y. Wang, P. Huar$_{r}$ e T. Kuo, 20.9 An energy-recycling three-switch singleinductor dual-input buck/boost DC-$_{2}$ converter with 93% 0.5mm^2 active area for light energy harvesting, in *2015 IEEE I⸳ ¡rnational Solid-State Circuits Conference - (ISSCC) Digest of Technical Papers* (2015), pp tc^{-3}

28. X. Liu, E. Sanchez-Sinenc$_z$ f An 86% efficiency 12 self-sustaining PV energy harvesting system with hysteresis regu⸳ $_{io}$on and time-domain MPPT for IoT smart nodes. IEEE J. Solid State Circuits **50**(6), 1424–1 A7 (2015)

29. B. Tar, U. Cilingiroglu, Na⸳ (6vatt-scale power management for on-chip photovoltaic energy harvesting beacons. IEEE J. H$_{H}$nerging Sel. Top. Circuits Syst. **4**(3), 284–291 (2014)

30. T. Ozaki, T. Hirose, H. As$_{ert}$, N. Kuroki, M. Numa, Fully-integrated highconversion-ratio dual-output voltage boost c$_{kin}$erter with MPPT for low-voltage energy harvesting. IEEE J. Solid State Circuits **51**(10), 12$_2$8–2407 (2016)

31. S. Carreon-Bautista, A. Ela$_g$, vy, A. Nader Mohieldin, E. Sanchez-Sinencio, Boost converter with dynamic input impedai$_{DC}$matching for energy harvesting with multi-array thermoelectric generators. IEEE Trans. Ind $_{ute}$lectron. **61**(10), 5345–5353 (2014)

32. G. Chowdary, A. Singh, S. ($_1$tterjee, An 18 na, 87% efficient solar, vibration and RF energy-harvesting power managen$_{io}$ t system with a single shared inductor. IEEE J. Solid State Circuits **51**(10), 2501–2513 $_{ati}$)16)

43⸳
⸀ov
En
⸀no
·on\
23⸳
da\

33. Y.K. Ramadass, A.P. Chandrakasan, An efficient piezoelectric energy harvesting interface circuit using a bias-flip rectifier and shared inductor. IEEE J. Solid State Circuits **45**(1), 189–204 (2010)

34. S. Bandyopadhyay, A.P. Chandrakasan, Platform architecture for solar, thermal, and vibration energy combining with MPPT and single inductor. IEEE J. Solid State Circuits **47**(9), 2199–2215 (2012)

35. G. Chowdary, S. Chatterjee, A 300 nw sensitive, 50 na DC-DC converter for energy harvesting applications. IEEE Trans. Circuits Syst. Regul. Pap. **62**(11), 2674–2684 (2015)

36. A. Devaraj, M. Megahed, Y. Liu, A. Ramachandran, T. Anand, A switched capacitor multiple input single output energy harvester (solar + piezo) achieving 74.6% efficiency with simultaneous MPPT. IEEE Trans. Circuits Syst. I Regl. Pap. 1–12 (2019)

37. T. Instruments, 6.4-GSPS single-channel or 3.2-GSPS dual-channel, 8-bit, RF-sampling analog-to-digital converter (ADC), ADC08DJ3200 (2020). https://www.ti.com/

38. A. Devices, 14-bit, 500 MSPS, JESD204B, quad analog-to-digital converter, AD9694 (2020). https://wwww.analog.com/

39. L. Technology, Constant-current DC/DC led driver in thinsot, LT1932 (2020). https://www.analog.com/

40. S. Bandyopadhyay, P.P. Mercier, A.C. Lysaght, K.M. Stankovic, A.P. Chandrakasan, A 1.1 nW energy-harvesting system with 544 pw quiescent power for next-generation implants. IEEE J. Solid State Circuits **49**(12), 2812–2824 (2014)

41. P. Weng, H. Tang, P. Ku, L. Lu, 50 mV input batteryless boost converter for thermal energy harvesting. IEEE J. Solid State Circuits **48**(4), 1031–1041 (2013)

42. A. Shrivastava, N.E. Roberts, O.U. Khan, D.D. Wentzloff, B.H. Calhoun, A 10 mV input boost converter with inductor peak current control and zero detection for thermoelectric and solar energy harvesting with 220 mV cold-start and -14.5 dBm, 915 MHz RF kick-start. IEEE J. Solid State Circuits **50**(8), 1820–1832 (2015)

43. E. Dallago, A. Lazzarini Barnabei, A. Liberale, G. Torelli, G. Venchi, A 300 mV low-power management system for energy harvesting applications. IEEE Trans. Power Electron. **31**(3), 2273–2281 (2016)

44. M.R. Elhebeary, M.A.A. Ibrahim, M.M. Aboudina, A.N. Mohieldin, Dual-source self-start high-efficiency microscale smart energy harvesting system for IoT. IEEE Trans. Ind. Electron. **65**(1), 342–351 (2018)

45. J. Katic, S. Rodriguez, A. Rusu, A high-efficiency energy harvesting interface for implanted biofuel cell and thermal harvesters. IEEE Trans. Power Electron. **33**(5), 4125–4134 (2018)

46. T. Paing, E.A. Falkenstein, R. Zane, Z. Popovic, Custom IC for ultralow power RF energy scavenging. IEEE Trans. Power Electron. **26**(6), 1620–1626 (2011)

47. T. Paing, J. Shin, R. Zane, Z. Popovic, Resistor emulation approach to low-power RF energy harvesting. IEEE Trans. Power Electron. **23**(3), 1494–1501 (2008)

48. D. El-Damak, A.P. Chandrakasan, A 10 nw – 1 μw power management IC with integrated battery management and self-startup for energy harvesting applications. IEEE J. Solid State Circuits **51**(4), 943–954 (2016)

49. V. Marian, B. Allard, C. Vollaire, J. Verdier, Strategy for microwave energy harvesting from ambient field or a feeding source. IEEE Trans. Power Electron. **27**(11), 4481–4491 (2012)

50. K. Kadirvel, Y. Ramadass, U. Lyles, J. Carpenter, V. Ivanov, V. McNeil, A. Chandrakasan, B. Lum-Shue-Chan, A 330nA energy-harvesting charger with battery management for solar and thermoelectric energy harvesting, in *2012 IEEE International Solid-State Circuits Conference* (2012), pp. 106–108

51. Y.K. Ramadass, A.P. Chandrakasan, A battery-less thermoelectric energy harvesting interface circuit with 35 mV startup voltage. IEEE J. Solid State Circuits **46**(1), 333–341 (2011)

52. P. Chen, C. Wu, K. Lin, A 50 nW-to-10 mW output power tri-mode digital buck converter with self-tracking zero current detection for photovoltaic energy harvesting. IEEE J. Solid State Circuits **51**(2), 523–532 (2016)

53. A. Richelli, S. Comensoli, Z.M. Kovacs-Vajna, A DC/DC boosting technique and power management for ultralow-voltage energy harvesting applications. IEEE Trans. Ind. Electron. **59**(6), 2701–2708 (2012)
54. M. Arrawatia, V. Diddi, H. Kochar, M.S. Baghini, G. Kumar, An integrated CMOS RF energy harvester with differential microstrip antenna and on-chip charger, in *2012 25th International Conference on VLSI Design* (2012), pp. 209–214
55. E.J. Carlson, K. Strunz, B.P. Otis, A 20 mV input boost converter with efficient digital control for thermoelectric energy harvesting. IEEE J. Solid State Circuits **45**(4), 741–750 (2010)
56. G. Papotto, F. Carrara, G. Palmisano, A 90-nM CMOS threshold-compensated RF energy harvester. IEEE J. Solid State Circuits **46**(9), 1985–1997 (2011)
57. M. Dini, A. Romani, M. Filippi, M. Tartagni, A nanocurrent power management IC for low-voltage energy harvesting sources. IEEE Trans. Power Electron. **31**(6), 4292–4304 (2016)
58. J. Im, S. Wang, S. Ryu, G. Cho, A 40 mV transformer-reuse self-startup boost converter with MPPT control for thermoelectric energy harvesting. IEEE J. Solid State Circuits **47**(12), 3055–3067 (2012)
59. X. Liu, E. Sanchez-Sinencio, A 0.45-to-3v reconfigurable charge-pump energy harvester with two-dimensional MPPT for internet of things, in *2015 IEEE International Solid-State Circuits Conference – (ISSCC) Digest of Technical Papers* (2015), pp. 1–3
60. H. Kim, S. Kim, C. Kwon, Y. Min, C. Kim, S. Kim, An energy-efficient fast maximum power point tracking circuit in an 800-μw photovoltaic energy harvester. IEEE Trans. Power Electron. **28**(6), 2927–2935 (2013)
61. G. Yu, K.W.R. Chew, Z.C. Sun, H. Tang, L. Siek, A 400 nW single-inductor dual input-tri-output DC-DC buck-boost converter with maximum power point tracking for indoor photovoltaic energy harvesting. IEEE J. Solid State Circuits **50**(11), 2758–2772 (2015)
62. C. Huang, Y. Ma, W. Yang, Y. Lin, C. Kuo, K. Chen, H. Liu, P. Yu, F. Chu, C. Lin, H. Huang, K. Hung, Y. Chu, Y. Lin, S. Kim, K. Ravichandran, A 99.2% tracking accuracy single-inductor quadruple-input-quadruple-output buck-boost converter topology with periodical interval perturbation and observation MPPT, in *2018 IEEE Asian SolidState Circuits Conference (A-SSCC)* (2018), pp. 171–174
63. S.B. Kjaer, Evaluation of the "hill climbing" and the "incremental conductance" maximum power point trackers for photovoltaic power systems. IEEE Trans. Energy Convers. **27**(4), 922–929 (2012)
64. M. S. et. al., A portable 2-transistor picowatt temperature-compensated voltage reference operating at 0.5 v. IEEE J. Solid State Circuits **47**(10), 2534–2545 (2012)
65. P. Chen, K. Ishida, K. Ikeuchi, X. Zhang, K. Honda, Y. Okuma, Y. Ryu, M. Takamiya, T. Sakurai, Startup techniques for 95 mV step-up converter by capacitor pass-on scheme and vth-tuned oscillator with fixed charge programming. IEEE J. Solid State Circuits **47**(5), 1252–1260 (2012)
66. P. Chen, X. Zhang, K. Ishida, Y. Okuma, Y. Ryu, M. Takamiya, T. Sakurai, An 80 mV startup dual-mode boost converter by charge-pumped pulse generator and threshold voltage tuned oscillator with hot carrier injection. IEEE J. Solid State Circuits **47**(11), 2554–2562 (2012)
67. J. Goeppert, Y. Manoli, Fully integrated start-up at 70 mV of boost converters for thermoelectric energy harvesting, in *ESSCIRC Conference 2015 – 41st European SolidState Circuits Conference (ESSCIRC)* (2015), pp. 233–236
68. P.M. Furth, Y.C. Tsen, V.B. Kulkarni, T.K.P.H. Raju, On the design of lowpower CMOS comparators with programmable hysteresis, in *2010 53rd IEEE International Midwest Symposium on Circuits and Systems* (2010), pp. 1077–1080
69. Coilcraft, Low profile shielded power inductors, LPS6235 (2018). http://www.coilcraft.com/lps6235.cfm#table
70. Murata, Chip monolithic ceramic capacitors, GRM21BF51C106ZE15l (2017)

Chapter 6
Sampling and Recovery of Signals with Spectral Sparsity

6.1 Introduction

Recent advances in flexible and wearable sensor based portable and wearable medical monitoring systems exhibit characteristics in the form of low-power consumption and high energy efficiency due to the long battery life requirements. Moreover, flexible bio-electronics systems manifest limited bandwidth/throughput due to the inherently large parasitic capacitance of the materials leading to limited sampling rate. With the traditional Nyquist sampling, this limit on the sampling rate results in (a) acquisition of bio-potential signals with restricted bandwidth thereby limiting the range of applications and (b) reduced number of parallel channel acquisition due to the reduced data rate support. For flexible bio-electronics based systems with adaptable architectures based on the application scenarios and constraints (viz., number of parallel channels, bandwidth of analog signal, power consumption, data-bit resolution, data transfer rate, data transmission power), it is pertinent to arrive at an efficient sampling rate that could be well below Nyquist. This problem leads us to innovate on sub-Nyquist sampling based data acquisition circuits and systems for flexible bio-electronic applications.

Further, in bio-telemetry the biomedical data must be compressed in time and space to reduce the memory requirements as well as reduce the transmission power to reduce the overall system power. Rapid growth in silicon miniaturization has led to intensive digital signal processing on-chip along with low-power analog-front-end realizations to realize complex flexible and non-flexible biomedical system with high energy efficiency. A brief overview of the sampling systems involving Nyquist and sub-Nyquist architectures is shown in Fig. 6.1. The Shannon–Nyquist sampling theorem has been the governing principle for data acquisition systems, which acts

Dr. Shuchin Aeron (Tufts University) and Dr. Laxmeesha Somappa (IIT Bombay) contributed to this chapter.

© The Author(s), under exclusive license to Springer Nature Switzerland AG 2022
S. Sonkusale et al., *Flexible Bioelectronics with Power Autonomous Sensing and Data Analytics*, https://doi.org/10.1007/978-3-030-98538-7_6

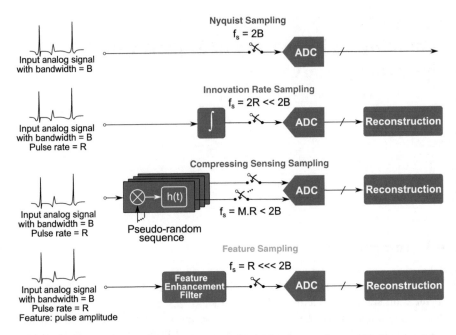

Fig. 6.1 Overview: Illustration of the Nyquist and sub-Nyquist sampling architectures involving interfacing of analog world of sensors with the digital world of signal processing

as the interface between the analog world of sensors and the digital world of processing. However, much of the medical monitoring applications involve signals whose physical bandwidth is much higher than the actual information rate. This a priori information of the signals can be exploited to realize complex biomedical systems that are extremely power and energy efficient. In principle, this a priori information leads to sub-Nyquist sampling of signals while still preserving the information that can then be recovered/reconstructed with digital signal processing. The subsequent chapters provide a peek into the theory and implementation of finite rate innovation sampling and compressed sensing based sampling.

In this chapter, we build from the basic Nyquist sampling leading up to sub-Nyquist sampling architectures including finite-rate-of-innovation (FRI) sampling. A brief overview of FRI sampling based application to pertinent biomedical data acquisition systems will be discussed.

6.2   Is Sampling at Nyquist Rate Necessary?

We begin by noting the well-known Shannon–Nyquist sampling theorem for a *real* bandlimited signal $x(t)$, with *one-sided* bandwidth B *hz*. The main result can be summarized as follows.

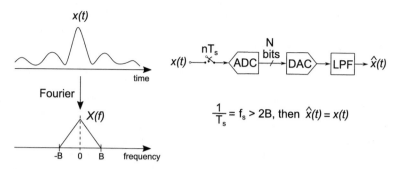

Fig. 6.2 Illustration of the Shannon–Nyquist sampling theorem

Theorem 1.1 *Given a real-valued signal* $x(t)$, *such that its Fourier Transform* $X(j2\pi f) = 0$ *for all* $f \geq B$ *hz, the signal can be recovered from its samples* $x(nT)$, $n \in \mathbb{Z}$, *if the sampling frequency satisfies* $\frac{1}{T} \geq 2B$ *hz. The optimal recovery is achieved via interpolation using an ideal low-pass filter with cutoff B Hz.*

This result is diagrammatically depicted in Fig. 6.2. From an operational point of view, we note some caveats here. The Analog to Digital Converter (ADC) is not infinite precision and therefore the measurements, at the very least, suffer from quantization noise.

For bandlimited systems, the frequency $2B$ is also referred to as the Nyquist frequency. A simple proof of this result can be found in any standard textbook [1]. Stated as such Theorem 1.1 does not take into account any additional structure that may be present in the data. To begin with, let us consider the simplest of the structure first, namely that the signal can be represented as a superposition of a finite number of damped or undamped exponentials, also known as the *Sum of Exponentials* (SoE) model [2].

$$y(t) = \sum_{m=1}^{M} c_m e^{-(\alpha_m - j2\pi f_m)t}. \tag{6.1}$$

Here $c_m \in \mathbb{C}$ are coefficients, and $f_m \in \mathbb{R}, \alpha_m \in \mathbb{R}$ are set of frequencies and attenuation factors, respectively. Let $B = \max\{f_1, \cdots, f_m\}$ be the maximum frequency. In this case we could apply Theorem 1.1, when $\alpha_m = 0$ for all m. Nevertheless, as we will show below, Nyquist sampling theorem is an overkill for this situation as it does not exploit the *finite information content* and the structure present in the SoE model.

We note that stated as such, in the SoE model, the signal is complex valued, as it is easy to explain the method with the complex valued case. In this case the measurements are also complex valued measurements. Another point to bear in mind is that the SoE model can be applied for sampling signals that are composed of superposition of finite number of impulses in the time domain, i.e., $x(t) = \sum_{m=1}^{M} c_m \delta(t - t_m)$. In this case, the measurements correspond to taking the Fourier transform of the signal and interchanging the role of fm and t in Eq. 6.1. We will come back to a more general case of this set-up when we consider sampling and recovery of Finite Rate of Innovation (FRI) signals.

For the SoE model, we have the following result, which is essentially distilled from [3].

Theorem 1.2 *Consider the model of Eq. 6.1. Then one can estimate the parameters for the SoE model from $N = 2M$ measurements $y(nT)$, $n = 0, 1, \cdots, 2M - 1$.*

The proof of this result is constructive and it is beneficial to go over it, as the basic principle will form the basis for recovery for another model, namely the Finite Rate of Innovation (FRI) signals that we will take up later. The essential idea is to exploit the particular exponential structure of the SoE model.

Let $z_m = e^{s_m T}$ with $s_m = -\alpha_m + j2\pi f_m$. Then one can rewrite the samples as

$$y(nT) = \sum_{m=1}^{M} c_m z_m^n, \tag{6.2}$$

for $n = 0, 1, \cdots, N - 1$. Equation 6.2 can be seen as a polynomial in the variables z_m. The method that achieves the result in Theorem 1.2 is based on what is called as the Matrix Pencil (MP) method [3]. It is also related to and similar to Prony's method and MUSIC [4]. MP begins by picking a parameter referred to as the *pencil parameter* ℓ that is typically chosen to be $\ell = \frac{N}{2}$. Letting $y(n) = x(nT)$, given the pencil parameter L, one forms two matrices $\mathbf{Y}_1, \mathbf{Y}_2$ as follows:

$$\mathbf{Y}_1 = \begin{bmatrix} y(0) & y(1) & \cdots & y(L-1) \\ y(1) & y(2) & \cdots & y(L) \\ \vdots & \vdots & & \vdots \\ y(N-L-1) & y(N-L) & \cdots & y(N-2) \end{bmatrix} \in \mathbb{C}^{N-L \times L} \tag{6.3}$$

$$\mathbf{Y}_2 = \begin{bmatrix} y(1) & y(2) & \cdots & y(L) \\ y(2) & y(3) & \cdots & y(L+1) \\ \vdots & \vdots & & \vdots \\ y(N-L) & y(N-L+1) & \cdots & y(N-1) \end{bmatrix} \in \mathbb{C}^{N-L \times L} \tag{6.4}$$

We note that these matrices are constant along the antidiagonals and are referred to as Hankel matrices. Introduce two other matrices:

$$\mathbf{Z}_1 = \begin{bmatrix} 1 & 1 & \cdots & 1 \\ z_1 & z_2 & \cdots & z_M \\ \vdots & \vdots & & \vdots \\ z_1^{N-L-1} & z_2^{N-L-1} & \cdots & z_M^{N-L-1} \end{bmatrix}, \mathbf{Z}_2 = \begin{bmatrix} 1 & 1 & \cdots & 1 \\ z_1 & z_2 & \cdots & z_M \\ \vdots & \vdots & & \vdots \\ z_1^{L-1} & z_2^{L-1} & \cdots & z_M^{L-1} \end{bmatrix}. \tag{6.5}$$

Then we see that

$$\mathbf{Y}_1 = \mathbf{Z}_1 \mathbf{C} \mathbf{Z}_2^\top, \quad \mathbf{Y}_2 = \mathbf{Z}_1 \mathbf{C} \mathbf{Z}_0 \mathbf{Z}_2^\top, \tag{6.6}$$

where $\mathbf{C} = \text{diag}[c_1, c_2, \ldots, c_M]$ and $\mathbf{Z}_0 = \text{diag}[z_1, z_2, \ldots, z_M]$ are $M \times M$ diagonal matrices. The main insight in the MP approach is that if $M \leq L \leq N - M$, the Matrix Pencil, $\mathbf{Y}_2 - \lambda \mathbf{Y}_1 = \mathbf{Z}_1 \mathbf{C}(\mathbf{Z}_0 - \lambda \mathbf{I})\mathbf{Z}_2^\top$ has rank M, and looses rank by one, for any $\lambda = z_m$.

In other words, for $M \leq L \leq N - M$, $z_1, \cdots, z_M$ are generalized eigenvalues of the pair of matrices $\mathbf{Y}_1, \mathbf{Y}_2$. Based on these observations, one may estimate $z = [z_1, \cdots, z_M]^\top$ via computing the eigenvalues of the matrix $\mathbf{Y}_1^\dagger \mathbf{Y}_2$, where $\mathbf{Y}_1^\dagger$ represents the Moore–Penrose pseudo-inverse of $\mathbf{Y}_1$. If the model order M is known in advance or an upper bound is known in advance, then one may select $L = M$, and $N = 2M$. However a bit of oversampling usually helps in presence of noise as the matrices $\mathbf{Y}_1, \mathbf{Y}_2$ tend to be highly *ill-conditioned*. This ill-conditioning also depends on the sampling period T. In presence of noise in the measurements, which may be inevitable when considering the quantization error, one may regularize and use truncated SVD based approach [3]. Finally, one notes that once z is estimated, the coefficients can be estimated by solving a simple linear system of equations.

When $\alpha_m = 0$, solving for the parameters of the SoE model of Eq. 6.1 is referred to as the problem of *line spectral estimation*, a problem that has received renewed interest recently [5, 6]. Detailed analyses of the MP method and its variants were carried out in [7, 8]. One of the main conclusions from [8] is that for the case of line spectral estimation, in presence of noise, the sampling period T and the number of samples N determine the limit of *super-resolution*, i.e., the ability to resolve two very close frequencies, of the MP method. Stated informally, when considering the *normalized case* where $f_m \in (0, 1]$ and with $T = 1$, for some small $\epsilon > 0$, if $N < \frac{1-\epsilon}{\delta}$, then MP method *cannot* resolve the frequencies with separation $\delta \leq \frac{1}{N}$ in presence of noise. On the other hand with $N \geq \frac{1}{\delta} + 1$, a slightly modified MP method is stable in presence of noise. In his context we will also like to point to a more recent work [9], where the authors point out the sensitivity of the MP method with respect to the sampling period T.

We will come back to the problem of line spectral estimation again in Chapter 2, where we introduce the framework of Compressed Sensing (CS), which is a flexible and generic framework for optimal sampling and recovery of information sparse signals. In the CS framework, one obtains similar limits to super-resolution. However, we would like to point out though that while the limits of super-resolution is well studied for the case when $\alpha_m = 0$, the authors are not aware of parallel results for the case when $\alpha_m \neq 0$. Finally, we would like to note that we made no attempts to exhaust the literature here and may have inadvertently missed referencing some works here from the ocean of literature on this topic out there.

In the next section we discuss the sampling and recovery of signals with finite rate of innovation and show how the sampling and recovery of such signals is related to the MP method introduced in this section.

6.3   Sampling Signals with Finite Rate of Innovation (FRI)

The study of signals with Finite Rate of innovation was initiated in [10] and followed up in a large body of work. The FRI signals are defined as follows:

Definition 1.1 ([10]) A signal $x(t) = \sum_{k \in \mathbb{Z}} c_k \varphi(t - \tau_k)$ for some base signal $\varphi(t)$, which in turn is parameterized via a finite number of parameters, is a signal with finite rate of innovation if

$$r = \lim_{T \to \infty} \frac{1}{T} C_x \left(\left[-\frac{T}{2}, \frac{T}{2} \right] \right) < \infty, \tag{6.7}$$

where C_x measures the degrees of freedom of the signal in the time window $[-\frac{T}{2}, \frac{T}{2}]$.

The main idea here is that in any finite window length say W seconds, degrees of freedom of the signal is on an average no more than $rW < \infty$.

To illustrate the approach behind sampling and recovery of the FRI signals, let us consider the case when the signal is a superposition of spikes/stream of diracs, i.e., $\varphi(t) = \delta(t)$—the Dirac delta function. The approach is diagrammatically illustrated in Fig. 6.3a. Let the filter $h(t)$ in Fig. 6.3b be $h(t) = e^{-\frac{t^2}{\sigma^2}}$. Then the filtered signal is given by:

$$y(t) = \sum_k c_k e^{-\frac{(t-\tau_k)^2}{2\sigma^2}} = \sum_k c_k e^{-\frac{\tau_k^2}{2\sigma^2}} e^{-\frac{t^2}{2\sigma^2}} e^{\frac{t\tau_k}{\sigma^2}}. \tag{6.8}$$

Now taking the sampling at intervals $nT, n = 0, 1, \ldots$, the sampled output is given by:

$$y[n] = \sum_k c_k e^{-\frac{\tau_k^2}{2\sigma^2}} e^{-\frac{(nT)^2}{2\sigma^2}} e^{\frac{nT\tau_k}{\sigma^2}}. \tag{6.9}$$

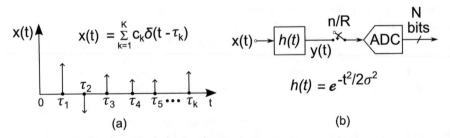

(a)   (b)

Fig. 6.3 For FRI signals, this can be thought of as an Analog to Information Converter (AIC), that is practically realizable, albeit with a high precision ADC

We note that if we let $c_k' = c_k e^{-\tau_k^2/2\sigma^2}$ and $z_k = e^{T\tau_k/\sigma^2}$, then with $y'[n] = y'e^{-\frac{(nT)^2}{2\sigma^2}}$, we have $y'[n] = \sum_k c_k' z_k^n$, which is an SoE model (compare with Eq. 6.1). Therefore, one may use MP and its variants to estimate the parameters of this model and using them to estimate c_k, τ_k. This approach can be applied to data in disjoint time windows of size W, assuming that under the FRI assumption, there are no more than $M = rW$ spikes. In the noiseless case, from Theorem 1.2 it follows that one can sample at a rate $R \geq 2r$ for accurate recovery. For analysis with another filter, namely a low-pass filter, hence making $h(t)$ a sinc function, we refer the reader to the seminal paper [10]. In a nutshell, it was shown in [10] that as long as $\varphi(t)$ is modeled using a finite number of parameters, say a piece wise polynomial or spline, the generic method and FRI sampling and recovery can be effectively used.

In Fig. 6.3, we call the basic architecture in the right panel an Analog to Information Converter (AIC) Architecture based on the FRI theory, in short the AIC-FRI architecture. An AIC can be realized using a conventional SAR (successive approximation register) ADC operating at sub-Nyquist rates, thereby combining the sub-Nyquist sampling and quantization. Sub-Nyquist sampling helps further reduce the dynamic power consumption and relax the linearity requirement of the front-end sample-and-hold circuit in the ADCs leading to highly energy-efficient data converters. Below we briefly survey several key extensions of this basic idea and approach that address operational considerations. Some of these aspects may be quite relevant to researchers in the biomedical domain for practically implementing the AIC architecture of Fig. 6.3. Note that below we have made no attempts to exhaust the literature on FRI sampling and recovery and its extensions.

6.3.1 Extensions and Operational Considerations

Apart from the practical issues of measurement noise and quantization, the reader may immediately ask if it is possible to implement the required filter, e.g., the exponential case considered above, with a physically realizable system, and how can one deal with more general case with φ not being delta functions. The issue of unknown $\varphi(t)$ was recently taken up in [11]. Toward addressing the filter issue, in the seminal work of [12] the main insight is to look for kernels (filters) with finite support that reproduce exponentials and polynomials. It turns out that many filters with rational Fourier transform satisfy these properties. Under this reproducible condition, one can mathematically reduce the set up to recovery of an SoE model from samples. Further, in [13], it was shown that as long as the filter satisfies the reproducing property approximately, it is still possible to use the FRI sampling and recovery in a stable manner (albeit with a linear transformation applied to the sampled data $y[n]$).

As pointed out in Sect. 6.2, the MP based approach remains sensitive to additive noise and it is also the case with the recovery methods that use the AIC-FRI architecture. To address this, oversampling is usually a good idea. Recently several methods have been proposed to deal with noise as well as the problem of model order selection, i.e., when the rate of innovation is not known. Among these, we would like to point out the model-fitting approach of [14, 15], and perhaps more recent work based on using Deep Neural Networks (DNNs) [16]. Although there has not been a significant amount of work that deals with various degrees of quantization of the measurements $y[n]$ in the AIC-FRI architectures, we believe that the recovery approach in [14, 15] can be easily extended and applied to this setting.

It is evident that Matrix Pencil (MP) and other related methods are dependent on *uniform sampling*, which may not be the case all the time. In order to alleviate this issue, [17] proposes using the idea of structured nonuniform sampling. In [18], the authors propose to use a hyperbolic secant kernel (filter) to deal with nonuniform sampling. Among more recent and possibly exciting avenues of making AIC-FRI architecture closer to hardware implementation; see [19].

We now briefly discuss where the ideas presented in this chapter have been applied in the context of biomedical signal processing.

6.4 Applications to Biomedical Signal Sampling and Recovery

The FRI sampling theory was used in [20] to effectively compress the ECG signal. The ECG was split into two parts: (a) the QRS complex is modelled using a nonuniform spline with K pieces and (b) the residual bandlimited signal with a bandwidth of L, as illustrated in Fig. 6.4. The ECG signal is then reconstructed using FRI sampling the QRS signal and separately reconstructing the ECG signal. On the contrary, [21] presents a Variable Pulse Width FRI (VPW-FRI) approach, where the entire ECG signal is sampled at the finite rate of innovation and then reconstructed. The VPW-FRI is an extension of the Dirac streams discussed in this section but generalized by adding two extra parameters in the form of *width* and *asymmetry* to the Dirac pulse.

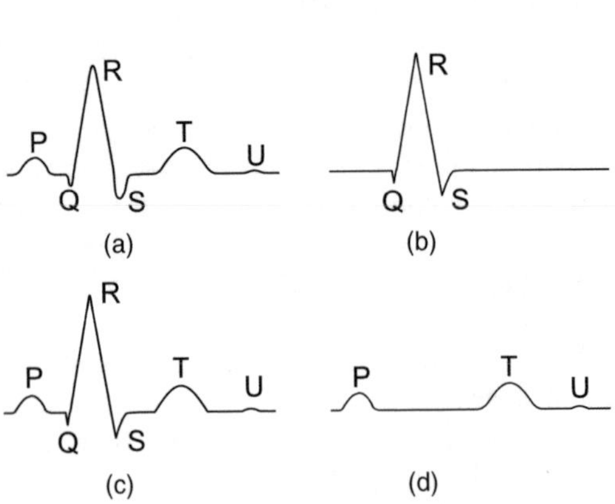

Fig. 6.4 (a) Original ECG signal, (b) the nonuniform spline approximation of the QRS complex, (c) the sum of the nonuniform spline and the bandlimited signal, and (d) the bandlimited approximation of the remaining part of the ECG signal

An approach similar to the ECG signal acquisition and reconstruction using FRI sampling was incorporated in [22] for neonatal EEG seizure signal compression. The EEG signal is modelled as periodic nonuniform linear spline with K pieces [22]. Neonatal EEG seizure signals comprise paroxysmal events that are abrupt trains of rhythmic spike waves that have distinct periods [22]. Traditionally, a 10-channel acquisition system with an ADC of 16-bit resolution and a Nyquist sampling of 200 Hz is employed to record the EEG for a 24-h period, resulting in a large memory and/or transmission power. To overcome this, the recorded EEG signal is pre-processed in two steps. First, the peaks and troughs of the original EEG signal is determined, yielding equivalent Dirac representation at these locations. Second, the stream of Dirac pulses are integrated to obtain a nonuniform linear spline representation of the original EEG signal. The final step is to obtain $2K + 1$ (Fig. 6.3) contiguous spectral values (continuous time Fourier series representation) from N samples and to determine the spectral values of the stream of Dirac pulses using the spectral values of the nonuniform splines. At the reconstruction, the location of the Dirac pulses is first determined from the spectral values, which is then integrated to obtain the original nonuniform spline representation of the neonatal EEG signal. Hence, neonatal EEG signal is closely represented using only $2K$ spectral coefficients with FRI.

A method for spontaneous brain activity detection in functional magnetic resonance imaging (fMRI) was developed in [23]. The fMRI signals are modelled as a convolution of a stream of Diracs (which is the innovation signal) and a hemodynamic response function (HRF). An FRI sampling kernel based on the HRF model can be designed that then allows to retrieve/reconstruct the innovation instants in continuous domain [23]. As a case study, the authors in [23] demonstrated a feasibility study for spontaneous brain activity detection using fMRI data based on FRI sampling using simulated noisy fMRI signals. A statistical parametric mapping software as used to generate the canonical HRF signal and the convolved input signal is sampled at the rate of innovation. Next, the iterative Cadzow denoising method was used to denoise the signal and the reconstruction of the innovation signal is performed.

FRI sampling based systems offer significant benefits in the biomedical imaging and radar applications. Prospective readers can refer to the following case studies for more generalized FRI applications: multi-path unknown time-delay estimation (well-known problems in under-water acoustics, and wireless communications) [24], radar applications [25], ultrasound imaging [26], wide band communication [25], and image super-resolution [27].

References

1. A.V. Oppenheim, A.S. Willsky, S.H. Nawab, G.M. Hernández et al., *Signals & Systems* (Pearson Educación, London, 1997)
2. T.K. Sarkar, O. Pereira, Using the matrix pencil method to estimate the parameters of a sum of complex exponentials. IEEE Antennas Propag. Mag. **37**(1), 48–55 (1995)

3. Y. Hua, T.K. Sarkar, Matrix pencil method for estimating parameters of exponentially damped/undamped sinusoids in noise. IEEE Trans. Acoust. Speech Signal Process. **38**(5), 814–824 (1990)
4. H. Krim, M. Viberg, Two decades of array signal processing research: the parametric approach. IEEE Signal Process. Mag. **13**(4), 67–94 (1996)
5. G. Tang, B.N. Bhaskar, B. Recht, Near minimax line spectral estimation. IEEE Trans. Inf. Theory **61**(1), 499–512 (2014)
6. R. Heckel, M. Soltanolkotabi, Generalized line spectral estimation via convex optimization. IEEE Trans. Inf. Theory **64**(6), 4001–4023 (2017)
7. W. Liao, A. Fannjiang, Music for single-snapshot spectral estimation: stability and super-resolution. Appl. Comput. Harmon. Anal. **40**(1), 33–67 (2016)
8. A. Moitra, Super-resolution, extremal functions and the condition number of Vandermonde matrices, in *Proceedings of the Forty-Seventh Annual ACM Symposium on Theory of Computing* (2015), pp. 821–830
9. D. Batenkov, A. Bhandari, T. Blu, Rethinking super-resolution: the bandwidth selection problem, in *ICASSP 2019 – 2019 IEEE International Conference on Acoustics, Speech and Signal Processing (ICASSP)* (2019), pp. 5087–5091
10. M. Vetterli, P. Marziliano, T. Blu, Sampling signals with finite rate of innovation. IEEE Trans. Signal Proces. **50**(6), 1417–1428 (2002)
11. Y. Zhang, P. L. Dragotti, Sampling streams of pulses with unknown shapes. IEEE Trans. Signal Process. **64**(20), 5450–5465 (2016)
12. P.L. Dragotti, M. Vetterli, T. Blu, Sampling moments and reconstructing signals of finite rate of innovation: shannon meets strang-fix. IEEE Trans. Signal Process. **55**(5), 1741–1757 (2007)
13. J.A. Urigüen, T. Blu, P.L. Dragotti, Fri sampling with arbitrary kernels. IEEE Trans. Signal Process. **61**(21), 5310–5323 (2013)
14. Z. Dogan, C. Gilliam, T. Blu, D. Van De Ville, Reconstruction of finite rate of innovation signals with model-fitting approach. IEEE Trans. Signal Process. **63**(22), 6024–6036 (2015)
15. H. Pan, T. Blu, M. Vetterli, Towards generalized FRI sampling with an application to source resolution in radio astronomy. IEEE Trans. Signal Process. **65**(4), 821–835 (2016)
16. V.C. Leung, J.-J. Huang, P. L. Dragotti, Reconstruction of FRI signals using deep neural network approaches, in *ICASSP 2020-2020 IEEE International Conference on Acoustics, Speech and Signal Processing (ICASSP)* (IEEE, Piscataway, 2020), pp. 5430–5434
17. S. Mulleti, C. S. Seelamantula, Periodic non-uniform sampling for FRI signals, in *2015 IEEE International Conference on Acoustics, Speech and Signal Processing (ICASSP)* (2015), pp. 5942–5946
18. X. Wei, B. Thierry, P. Dragotti, Finite rate of innovation with non-uniform samples, in *2012 IEEE International Conference on Signal Processing, Communication and Computing (ICSPCC 2012)* (2012), pp. 369–372
19. S. Rudresh, A.J. Kamath, C. Sekhar Seelamantula, A time-based sampling framework for finite-rate-of-innovation signals, in *ICASSP 2020 – 2020 IEEE International Conference on Acoustics, Speech and Signal Processing (ICASSP)* (2020), pp. 5585–5589
20. Y. Hao, P. Marziliano, M. Vetterli, T. Blu, Compression of ECG as a signal with finite rate of innovation, in *2005 IEEE Engineering in Medicine and Biology 27th Annual Conference* (2005), pp. 7564–7567
21. G. Baechler, N. Freris, R.F. Quick, R.E. Crochiere, Finite rate of innovation based modeling and compression of ECG signals, in *2013 IEEE International Conference on Acoustics, Speech and Signal Processing* (2013), pp. 1252–1256
22. K.-K. Poh, P. Marziliano, Compression of neonatal EEG seizure signals with finite rate of innovation, in *2008 IEEE International Conference on Acoustics, Speech and Signal Processing* (2008), pp. 433–436
23. Z. Doğan, T. Blu, D. Van De Ville, Detecting spontaneous brain activity in functional magnetic resonance imaging using finite rate of innovation, in *2014 IEEE 11th International Symposium on Biomedical Imaging (ISBI)* (2014), pp. 1047–1050

24. K. Gedalyahu, Y.C. Eldar, Time-delay estimation from low-rate samples: a union of subspaces approach. IEEE Trans. Signal Process. **58**(6), 3017–3031 (2010)
25. W.U. Bajwa, K. Gedalyahu, Y.C. Eldar, Identification of parametric underspread linear systems and super-resolution radar. IEEE Trans. Signal Process. **59**(6), 2548–2561 (2011)
26. R. Tur, Y.C. Eldar, Z. Friedman, Innovation rate sampling of pulse streams with application to ultrasound imaging. IEEE Trans. Signal Process. **59**(4), 1827–1842 (2011)
27. I. Maravic, M. Vetterli, K. Ramchandran, Channel estimation and synchronization with sub-Nyquist sampling and application to ultra-wideband systems, in *2004 IEEE International Symposium on Circuits and Systems (IEEE Cat. No.04CH37512)*, vol. 5 (2004), pp. V–V

Chapter 7
Compressed Sensing

7.1 Introduction

In this chapter we introduce the compressed sensing (CS) theory for computation-ally feasible signal recording and reconstruction systems. This is followed by the application of CS in realizing analog-to-information converter circuits and systems for a wide range of emerging applications; particularly, hardware realizations for efficient biomedical data acquisition using flexible and traditional circuits based on CS are explored. The applications reported in the chapters are in no way extensive, but a subset highlighting contrasting implementations of sub-Nyquist rate circuits and systems for efficient data acquisition. Sub-Nyquist sampling leads to highly energy and power efficient design of analog-to-digital converter circuits together with orders of magnitude reduction in the energy cost of wireless data transmission. While a plethora of literature on CS and its applications can be found, we refer the readers to the most relevant references throughout the chapter.

7.2 What Is Compressed Sensing?

To relate and build off the development from Chap. 1, let us consider the case where a finite dimensional signal $x \in \mathbb{R}^n$ can be written as $x = Fc$, where F is the discrete Fourier transform (DFT) matrix. Suppose that c is *sparse*, i.e. the number of non-zero elements in c, say k satisfies $k << n$.

This can be thought of as a special case of the Sum of Exponential (SoE) model considered in the previous chapter, where the frequencies take values in a finite

Dr. Shuchin Aeron (Tufts University) and Dr. Laxmeesha Somappa (IIT Bombay) contributed to this chapter.

set of n frequencies $f_i = \frac{j2\pi i}{n}, i = 0, 1, \ldots, (n - 1)$ and hence the signal is a superposition of a small number of these frequencies. We note that in this setting one can also consider the setting where $x = \mathbf{F}^* c$, which corresponds to the signals that are sparse in (discrete) time. We already know for this setting that taking regularly spaced temporal or filtered samples is sufficient. However, we would like to generalize this setting significantly and go beyond the type of signal models considered, where $\mathbf{F}$ can be replaced with any basis, not necessarily Fourier.

Suppose now that one is allowed to make *linear* measurements of the type $y = a^\top x$. The problem of (noiseless) compressed sensing (CS) addresses the following question:

What are good classes of and minimal number of *linear measurements*, of the type $y_i = a_i^\top x, i = 1, 2, \cdots , m$ that are needed to recover a k-sparse signal x from measurements y in a *computationally feasible manner*?

The point that we would like to stress here is that computationally feasible recovery is a central aspect of the theory of CS. Removing that criteria leads to trivial conclusions in that starting with basic linear algebraic arguments one can show that $m = 2k$ measurements, collected as $y = \mathbf{A}x$, are necessary and sufficient for unique recovery of a k-sparse signal x, for any $\mathbf{A}$ with column rank $2k$. For a quick proof and insight, we refer the reader to [1]. But the algorithm for recovery can be seen to have a combinatorial complexity, in that one would have to solve $\binom{n}{k}$ linear sub-systems to figure out the k-sparse x. It is noteworthy that even if the values of non-zero locations of x are known, the number of measures to just figure out the right support is still $2k$. Thus, the main sample complexity arises from *uncertainty in the support* of x. The reader may now appreciate the name compressed sensing, in that although the latent dimension of x is n, if the signal is sparse, one can, in a completely *non-adaptive* manner, collect $2k$ linear measurements and recover x.

Let us now come to the *computationally feasible* part of compressed sensing. Essentially the question is how one can alleviate the computational complexity in recovery from these measurements. Informally speaking, one can appreciate that there is a tradeoff between the number of measurements and the computational complexity of the recovery. So the question really is what is that best tradeoff or nearly optimal tradeoff?

7.3 $\ell_1 - \ell_0$ Equivalence

Given the measurements y and the sending matrix $\mathbf{A}$, one can pose the following optimization problem to recover the signal:

$$\arg \min \|x\|_0 \text{ s.t. } y = \mathbf{A}x, \tag{7.1}$$

where $\|x\|_0$ is called as the ℓ_0 pseudo-norm and is equal to the number of non-zero entries in x. This problem is known to be NP-hard, i.e., largely computationally

Fig. 7.1 Different ℓ_p-balls in two-dimensional space

infeasible, in general [1]. Borrowing ideas from *convex relaxations* of integer and combinatorial optimization problems, the central approach in addressing the computationally feasible part of the CS framework is to consider the following optimization problem:

$$\arg\min \|x\|_1 \quad \text{s.t.} \quad y = Ax, \tag{7.2}$$

where $x\|_1$ denotes the ℓ_1-norm. To get a sense of different ℓ_p norms, please see Fig. 7.1. Being a linear program, this problem is computationally feasible. A question that arises naturally here is—Under what conditions on A, k, n does the solution of 7.1 coincide with 7.2? This phenomena is broadly referred to as the $\ell_1 - \ell_0$ equivalence. In this context, there has been an immense level of activity and a plethora of conditions and scenarios have been studied and characterized. Notable among these are—(a) Restricted Isometry Property (RIP) [2], (b) null-space property [3], and (c) incoherency and coherency [4, 5]—conditions on the measurement matrix A, to name a few. It is infeasible to provide an overview of all of these results here. We single out one approach, that is largely *geometric* in nature and from our perspective conveys the core ideas in a simple and crisp manner, which we outline below.

The following material is distilled completely from the notes and book by R. Vershynin [6, 7]. We should apologize in advance if we miss giving credit to other related work and if we have inadvertently missed referring to recent advances in this context. Without cluttering the discussion and to convey *essential* insights, we will present series of results without proofs and then stitch them together to arrive at conditions for $\ell_0 - \ell_1$ equivalence. In the process we will identify a number of ensembles for sensing matrices. While we will not provide complete mathematical details, the accompanying figures should convey all the geometrical intuition needed to appreciate the results.

Geometry [See Fig. 7.2]—Let $\mathcal{B}_1$ be the unit ℓ_1 ball, and let $C = \|x_0\|_1 \mathcal{B}_1$, which is nothing but a scaling of the unit ℓ_1 ball by $\|x_0\|_1$, where x_0 is the *true* signal. At any point $x \in C$, one defines a spherical section $\mathcal{S}(C, x)$ via

$$\mathcal{S}(C, x) = \left\{ \frac{z - x}{\|z - x\|_2}, z \in C \right\} \tag{7.3}$$

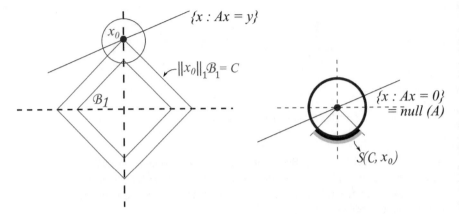

Fig. 7.2 Geometry of ℓ_1 minimization for exact recovery (left): Unit ℓ_1 is expanded/shrunk so that it interacts with the affine subspace $\{x : \mathbf{A}x = y\}$. Picture translated to the origin (right): for exact recovery, the subspace $\text{null}(\mathbf{A})$ should not intersect with the spherical section $mcS(C, x_0)$

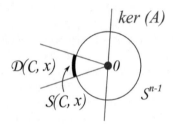

Fig. 7.3 Exact recovery condition illustration

Lemma 2.1 (Descent Cone Condition for Exact Recovery) *[6] A sufficient condition for exact recovery by solving 7.2 is*

$$S(C, x_0) \cap \text{null}(\mathbf{A}) = \varnothing. \tag{7.4}$$

The situation is depicted in Fig. 7.3, which is adapted from [6]. One can appreciate that the condition for recovery in Lemma 2.1 is essentially saying that as long as null space of $\mathbf{A}$ does not intersect with $S(C, x_0)$, exact recovery is possible. Clearly if a large number of measurements are taken, the null space of $\mathbf{A}$ is smaller in dimension, the *chances* are small that it will intersect a spherical cap of a certain size. On the other hand if the number of measurements are small, the null space of $\mathbf{A}$ is *large* and the chances of intersection are more. This intuition is made more precise in the following result, due to Gordon. In order to appreciate the result one needs the notion of *Gaussian width*, denoted by $w_G(S)$ of a set S. It is defined as

$$w_G(S) = \mathbb{E}_{g \sim \mathcal{N}(0, \mathbf{I}_n)} \left[\sup_{x \in S} g^\top x \right] \tag{7.5}$$

Fig. 7.4 Illustration of visualizing width

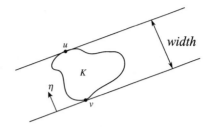

An insight into this quantity is provided in Fig. 7.4, which actually depicts a related quantity called as the spherical width defined as [7], $w_s = \mathbb{E}_{\eta \sim \mathsf{Unif}(\mathscr{S}^{n-1})} \sup_{x \in S} \eta^\top x$, where $\mathsf{Unif}(\mathscr{S}^{n-1})$ denotes the uniform distribution on the unit sphere $\mathscr{S}^{n-1}$. It turns out that these two quantities are within a scaling factor of $\sqrt{n}$ from each other [7].

Theorem 2.1 (Escape Theorem) *[7] Consider a set S of the surface of the unit sphere in n dimensions. Let $\mathbf{A} \in \mathbb{R}^{m \times n}$ be a matrix, whose rows are independent, isotropic, and sub-Gaussian random vectors in $\mathbb{R}^n$. If $m \geq C K^4 (w_G(S))^2$, then the random subspace $\mathsf{null}(A)$ satisfies*

$$S \cap \mathsf{null}(A) = \varnothing \tag{7.6}$$

with probability at least $1 - 2e^{-\frac{cm}{K^4}}$. Here K is a constant that only depends on the sub-Gaussianity of the distribution from which the rows of $\mathbf{A}$ are sampled.

We now need one more result before we can establish the $\ell_0 - \ell_1$ equivalence that we set out to do in the beginning. We need an estimate of the Gaussian width of $S(C, x_0)$ for any k-sparse vector x_0. This result appears in several papers, but we refer the reader to [8].

Lemma 2.2 *There exists a positive constant c such that, for all k-sparse x_0,*

$$S(C, x_0) \leq c\sqrt{k \log(2n/k)} \tag{7.7}$$

We are now ready to state and appreciate the following result.

Theorem 2.2 ($\ell_0 - \ell_1$ Equivalence for CS) *For the problem of compressed sensing (CS), if the rows of sensing matrix $\mathbf{A}$ are independent, isotropic, sub-Gaussian random vectors in $\mathbb{R}^n$, then for some constants c_1, c_2, if*

$$m \geq 2c_1 k \log(2n/k),$$

then with probability exceeding $1 - c_2 e^{-m}$, the solution to 7.1 is the same as the solution to 7.2.

It turns out that a number of ensembles satisfy the conditions of Theorem 2.2, namely sub-Gaussianity, needed for exact recovery of k-sparse vectors using ℓ_1 minimization. Primary examples are random Rademacher matrices with each entry i.i.d. Bernoulli $\{+1, -1\}$ with parameter $1/2$ and random Gaussian matrices. For more examples, we refer the reader to [9].

7.4 Compressed Sensing with Analysis Sparsity

We will close this section with discussion of a variant of the CS problem considered above, called as the analysis of sparse sensing and recovery problem [10, 11]. The main idea is as follows. Suppose there is an overcomplete dictionary (not forming an orthonormal basis) $\Omega \in \mathbb{R}^{p \times n}$, $p > n$ such that the signal x exhibits sparsity when analyzed through this dictionary, i.e., the vector of coefficients $c = \Omega x$ is sparse. Again following the same methodology of considering the convex relaxation of

$$\arg\min_x \|\Omega x\|_0 \quad \text{s.t.} \quad y = Ax, \tag{7.8}$$

to the following optimization problem:

$$\arg\min_x \|\Omega x\|_1 \quad \text{s.t.} \quad y = Ax. \tag{7.9}$$

Naturally, one may ask the same question as in the *synthesis* CS set-up considered before—when do the solutions to the two optimization problems 7.8 and 7.9 coincide under sparsity assumptions on the true Ωx? We will not dwell into this question here but will refer the readers to [10, 11].

7.5 Algorithms for Solving the ℓ_1 Minimization Problem

There have been a number of algorithms that have been proposed for large-scale compressive sensing problems. We only mention a few of them here by name and without getting into any detail. Some of the methods are Fast Iterative Shrinkage-Thresholding Algorithm (FISTA) [12], Orthogonal Matching Pursuit [13], Compressive Sampling Matching Pursuit (Co-SAMP) [14], and Iterative Hard Thresholding [1]. In the next section we review the hardware architectures for efficient sampling and recovery that are inspired by the theory and methods of compressive sensing.

7.6 Analog-to-Information Converters Based on CS

Sensing applications involving medical monitoring, radar applications, and sensor interface for IoT applications involve sensing signals with physical bandwidth higher than the actual information rate. Alternative sampling and digitization technique termed *Analog-to-Information Converters* (AIC) was introduced to take into account the a priori information in the form of sparseness in the sensing signal. This allows the signal to be sampled well below the Nyquist rate leading to efficient hardware and data transmission while still preserving the complete information. Figure 7.5 shows AIC based on CS that have been popular and relevant to practical applications and amenable to implementation on hardware. Analog

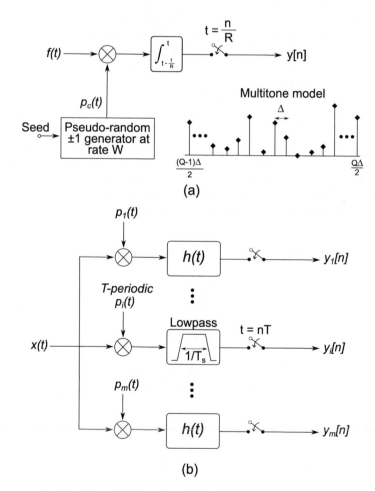

Fig. 7.5 (a) Random demodulator and (b) wide band converter architecture

signal compression is realized with two distinct AIC architectures, namely random demodulator (RD) and modulated wide band converter (MWBC).

The RD based CS-AIC was proposed [15] to acquire sparse, bandlimited signals in applications where the discrete number of frequency components of interest is much smaller than the band limit allows. The idea behind the RD AIC based on CS is illustrated in Fig. 7.5a. A high data rate pseudorandom chipping sequence is mixed (demodulated) with the input signal and the output is integrated and dumped at a constant rate R, which is smaller than the Nyquist rate. The digitization then requires a very low-data rate ADC leading to a low-power system realization. The signal reconstruction is performed using the sampled vector Y using CS techniques including ℓ_1 minimization.

While the RD AIC is suitable for applications involving sum of tones, it is not suitable for applications with multi-band signals. The MWBC system was proposed for multi-band system AIC realization in [16]. The system comprises of parallel front-end channels with each channel comprising a multiplier and a low-pass filter as demonstrated in Fig. 7.5b. To understand the operation for the MWBC AIC, we refer to Fig. 7.6. A multi-band signal $x(t)$ has sparse frequency bands (N bands) with each band limited to a bandwidth of B Hz, and a maximum frequency of f_{max} Hz. Due to the mixing and integration in each channel, the spectrum bands of input $x(t)$ are overlaid in the spectrum of the output $y[n]$. It must be noted that a single band in the input $x(t)$ occupies only two adjacent spectrum slices at the output, since the $T \geq B$ (refer to Figs. 7.5b and 7.6). For example, in Fig. 7.6, channels i and i' output spectrum slices will have a linear combination of the input spectrum bands with each input spectrum band occupying a maximum of two spectrum slices at the output. The reconstruction of the signal bands in $x(t)$ involves re-positioning of the slices of the output spectrum. An efficient approach termed as continuous to

Fig. 7.6 Illustration of the wide band converter operation with two channels

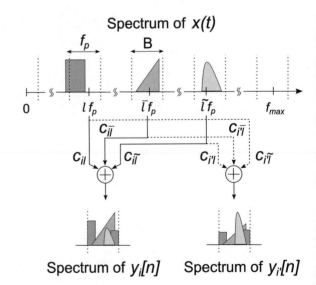

Spectrum of $y_i[n]$ Spectrum of $y_{i'}[n]$

finite (CTF) was proposed for reconstruction and we direct the readers to refer to [16] for a detailed description of the reconstruction signal processing hardware for the MWBC system. In the subsequent section we discuss flavors of RD and MWBC based AIC hardware realizations particularly for biomedical applications.

7.7 Application to Wearable and Flexible Biomedical Circuits and Systems

Biomedical applications involving wearable sensors are increasingly employed in medical monitoring. One such system implemented with a traditional Nyquist rate ADC is shown in Fig. 7.7a. The conventional system includes a low-noise amplifier (LNA) to efficiently acquire the bio-potential signal in the presence of noise, a Nyquist rate analog-to-digital converter (ADC) with a resolution of at-least 8 bits followed by a RF radio for wireless telemetry. At the receiver end, the data is received and the bio-potential signal is either recorded or monitored continuously. The main bottleneck in the traditional system arises from the fact that the energy cost of wireless data transmission is orders of magnitude greater than the other circuit blocks in the system. This high energy cost associated with the system suggests that some form of data compression needs to be performed at the transmission side to minimize the energy cost of the overall system, specifically in implantable applications such as multi-channel neural recordings.

In addition to low-power techniques for the front-end and ADC circuit implementation discussed in Chap. 4, the system must also encompass compressed sensing (CS) to reduce the amount of data to be processed and transmitted at the receiver side, to realize an energy-efficient bio-telemetry system. The CS techniques discussed in this section can be used to implement such systems as shown in Fig. 7.7b.

One such implementation of the CS based analog front end (AFE) for biomedical applications is shown in Fig. 7.9 [17]. The measurement matrix $[Y]$ is computed from the matrix multiplication of the random matrix $[\Phi]$ with the sampled bio-potential signal after the LNA, $[X]$. The M partial products are generated based

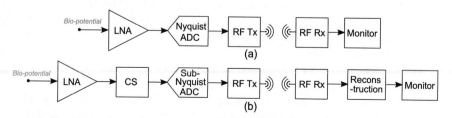

Fig. 7.7 Block-level description of bio-potential telemetry with (**a**) conventional approach and (**b**) compressed sensing based approach

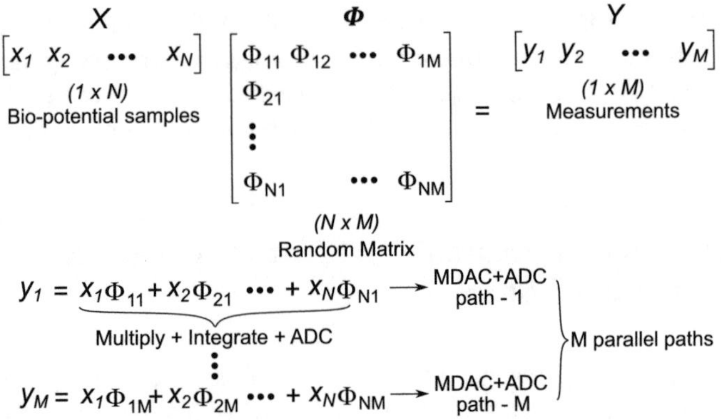

$$
\begin{array}{ccc}
X & \Phi & Y \\
\begin{bmatrix} x_1 & x_2 & \cdots & x_N \end{bmatrix} &
\begin{bmatrix}
\Phi_{11} & \Phi_{12} & \cdots & \Phi_{1M} \\
\Phi_{21} & & & \\
\vdots & & & \\
\Phi_{N1} & & \cdots & \Phi_{NM}
\end{bmatrix} &
\begin{bmatrix} y_1 & y_2 & \cdots & y_M \end{bmatrix} \\
\textit{(1 x N)} & & \textit{(1 x M)} \\
\text{Bio-potential samples} & = & \text{Measurements}
\end{array}
$$

(N x M)
Random Matrix

$$y_1 = \underbrace{x_1\Phi_{11} + x_2\Phi_{21} \cdots + x_N\Phi_{N1}}_{\text{Multiply + Integrate + ADC}} \rightarrow \begin{array}{c}\text{MDAC+ADC}\\\text{path - 1}\end{array}$$

$$\vdots$$

$$y_M = x_1\Phi_{1M} + x_2\Phi_{2M} \cdots + x_N\Phi_{NM} \rightarrow \begin{array}{c}\text{MDAC+ADC}\\\text{path - M}\end{array}$$

$\Big\}$ M parallel paths

Fig. 7.8 Operation of the CS-AFE with the MDAC and ADC

on the N input samples by first multiplying the input samples with the random generator, and integrating the successive N products of the multiplier and finally followed by one ADC operation of the integrated value to obtain the final compressed measurement value $[Y]$ (Fig. 7.8).

Each component y_i of the compressed measurement matrix $[Y]$ is computed using a switched capacitor (SC) based multiplying digital-to-analog-converter and integrator (MDAC-I) as shown in Fig. 7.9a. The MDAC-I in each of the M paths operate at the Nyquist rate f_s, and the subsequent 10-bit SAR ADC operates at a sub-Nyquist rate of $N_{SAR}f_s/N$, where N_{SAR} is the number of cycles needed for the SAR operation. The important advantage of the architecture is the ability to operate in a pipelined fashion, i.e., when the MDAC-I is operating on the present set of N input samples, the SAR ADC performs the digitization of the integrated value of the previous N cycles, thereby allowing for continuous sampling of the input signal and offering reduced hardware complexity.

The MDAC-I implementation shown in Fig. 7.9b operates in two clock phases, namely the sampling and multiplying phase ϕ_1 followed by the integrating phase ϕ_2. The input sampling capacitance is chosen based on the input-referred noise specification. Implementing a binary DAC will introduce stringent matching requirements leading to an additional calibration hardware [17]. In order to avoid this, a C-2C DAC array is implemented with relaxed matching requirements without any additional calibration hardware. Similarly for the SAR ADC implementation, a C-2C DAC approach can be used to minimize the area and power dissipation as against a binary capacitive DAC [17].

The random matrix $[\Phi]$ is generated using an on-chip linear feedback shift register (LFSR). Referring to Fig. 7.9a, the LFSR generates a 6-bit random number for each multiplication path. The multiplier hence has to implement a 6-bit multiplication with the incoming sampled analog data. This is achieved using a C-2C capacitive DAC implementation where each of the binary switches b_i $(i > 1)$ is controlled directly by the random bit r_i of the 6b-bit random number during the

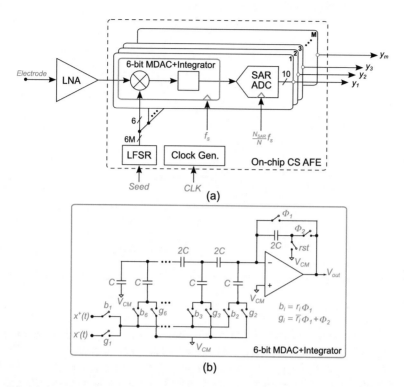

Fig. 7.9 (**a**) A compressed sensing based biomedical front end implemented using a multiplying DAC (MDAC), and (**b**) the on-chip multiplying DAC and integrator implemented using switched capacitor C-2C structure

multiplying phase ϕ_1. The MSB bit (r_1) acts as a sign-bit and switches the analog input sign to the subsequent switches based on the digital sign-bit. Further, a one-bit random modulator can also be supported by simply turning-on all the switches b_2 to b_6. At the reconstruction side, a **Fourier** basis is used and CVX $\ell_1 - norm$ convex optimization is used for the reconstruction [17].

One approach to reduce the required hardware and hence the power consumption while still achieving the required resolution in a CS based AFE is to use a segmented CS as shown in Fig. 7.10 [18]. In segmented CS, the random matrix Φ with dimensions $M \times N$ is adopted such that half the entries of the matrix are set to zero. Unlike the unsegmented case, every successive $N/2$ input samples of one frame share the same hardware to generate M measurement digital samples using only $M/2$ hardware paths as illustrated in Fig. 7.10. Briefly, the technique exploits hardware sharing with time multiplexing due to the inherent implementation of the segmented CS. Measurement samples y_1 to $y_{M/2}$ are generated first by the $M/2$ parallel hardware paths by multiplying with a random signal ϕ_A. Next, the input samples $N/2 + 1$ to N of the same frame are multiplied with the random signal ϕ_B to generate the remaining measurement samples $y_{M/2+1}$ to y_M.

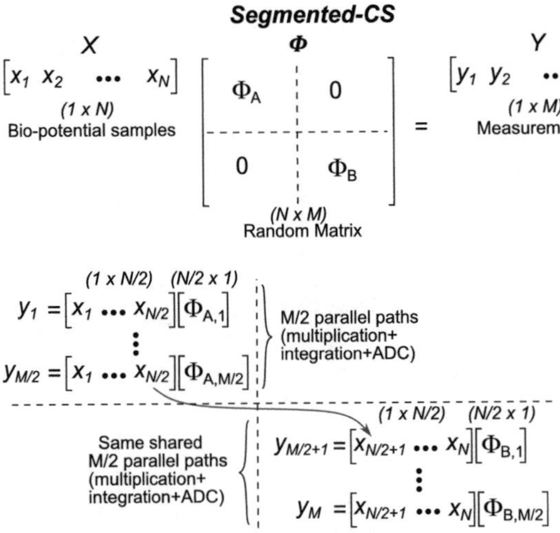

Segmented-CS

$$
\begin{bmatrix} x_1 & x_2 & \cdots & x_N \end{bmatrix} \quad \begin{bmatrix} \Phi_A & \vdots & 0 \\ \hdashline 0 & \vdots & \Phi_B \end{bmatrix} = \begin{bmatrix} y_1 & y_2 & \cdots & y_M \end{bmatrix}
$$

X (1 x N) Bio-potential samples

Φ (N x M) Random Matrix

Y (1 x M) Measurements

$$
y_1 = \begin{bmatrix} x_1 & \cdots & x_{N/2} \end{bmatrix} \begin{bmatrix} \Phi_{A,1} \end{bmatrix}
$$
$$
\vdots
$$
$$
y_{M/2} = \begin{bmatrix} x_1 & \cdots & x_{N/2} \end{bmatrix} \begin{bmatrix} \Phi_{A,M/2} \end{bmatrix}
$$

(1 x N/2) (N/2 x 1)

M/2 parallel paths (multiplication+ integration+ADC)

Same shared M/2 parallel paths (multiplication+ integration+ADC)

(1 x N/2) (N/2 x 1)

$$
y_{M/2+1} = \begin{bmatrix} x_{N/2+1} & \cdots & x_N \end{bmatrix} \begin{bmatrix} \Phi_{B,1} \end{bmatrix}
$$
$$
\vdots
$$
$$
y_M = \begin{bmatrix} x_{N/2+1} & \cdots & x_N \end{bmatrix} \begin{bmatrix} \Phi_{B,M/2} \end{bmatrix}
$$

Fig. 7.10 Operation of a segmented CS based analog front end with shared hardware for energy-efficient implementation

An analog front end with segmented compressed sensing for ECG monitoring with $M = 200$ was implemented using only 100 parallel paths in [18]. Unlike the implementation in Fig. 7.9, the technique adopted in Fig. 7.11 allows for the multiplication, integration, and the ADC operation into a single hardware using an incremental $\Delta\Sigma$ ADC. The CS-AFE discussed in Fig. 7.9a integrates all the input samples after which the ADC digitizes the integrated value, which poses a challenge in limited voltage headroom applications. The incremental $\Delta\Sigma$ ADC technique avoids this issue and allows the CS-AFE implementation at lower supply voltages. A sample-and-hold amplifier (SHA) and a capacitively coupled chopper instrumentation amplifier (IA) are shared among all the paths. The first order incremental $\Delta\Sigma$ ADC is realized using a switched capacitor integrator, DAC, and an up-down counter. The differential-to-single-ended (D2S) conversion and multiplication is achieved by switching the differential input signal to the sampling capacitor C_1, based on the sign of the pseudo-random-bit sequence ϕ_{RNG}. When the integrator output crosses the threshold voltage set by the references V_{REFP} and V_{REFN}, the counter updates and the logic controlling the DAC controls the appropriate switch to increase or decrease the integrator charge. The integrator is reset after every $N/2$ cycles since the next set of samples from index $N/2 + 1$ to N will use the same hardware as discussed earlier. To reduce the power consumption of the $M/2$ incremental ADCs, dynamic biasing is utilized to lower the current of the integrator Op-Amp during the sampling phase. Finally, at the receiver end a **sym8 wavelet** basis is used for the reconstruction of the acquired ECG signal [18].

Compressed sensing based bio-potential signal acquisition finds a useful application in wearable thin-film transistor (TFT) based circuits due to the inherent speed limitations (or unity gain frequency, f_T) associated with the high parasitic

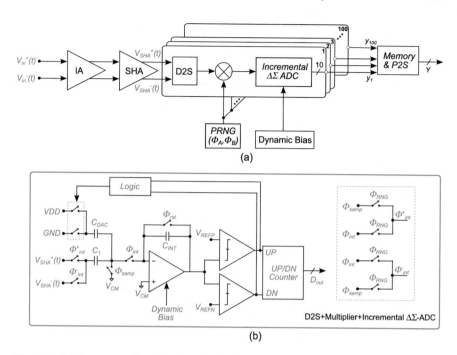

Fig. 7.11 (**a**) A compressed sensing based analog front end for ECG monitoring using segmented CS, and (**b**) the on-chip differential-to-single-ended (D2S) converter, multiplier and incremental $\Delta\Sigma$ ADC

capacitance, and absence of PMOS transistors in TFT technology [19]. As an example, for an EEG acquisition system with multiple channels, traditional non-flexible implementations pose challenge due to the large amount of cabling distributed over the patient's scalp, which impacts the mechanical stability of the electrodes and causes motion artifacts in the recording. To address this limitation, flexible electrode arrays with active circuits implemented on them for instrumentation can be thought of with minimal cables for EEG signal acquisition and processing. Further, the number of acquisition channels with the traditional Nyquist rate sampling is limited in TFT due to the large time constants and parasitic capacitances associated with the technology [19]. While a hybrid system involving TFT and CMOS circuits involve much of the complex functionalities implemented in CMOS, the low-noise amplifiers must be placed close to the electrodes and hence must be implemented on TFT. This reduces the susceptibility of the signal due to the otherwise interconnect being present between the electrodes and CMOS. Further, the number of interconnects to the CMOS can be reduced by multiplexing the channels onto a single interconnect to the CMOS.

A TFT based analog front end based on CS for a multi-channel EEG acquisition is shown in Fig. 7.12 [19]. The TFT circuit comprises of a chopper based LNA, a multiplier, and a passive integrator followed by a sub-Nyquist sampler, which is multiplexed between all the acquisition channels. This single interconnect is

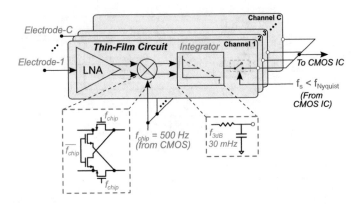

Fig. 7.12 Compressed sensing front-end circuit implementation based on thin-film transistors for wearable EEG acquisition

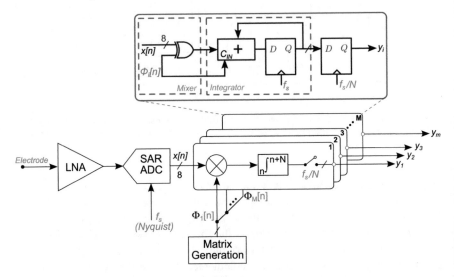

Fig. 7.13 An all-digital compressed sensing architecture for wireless bio-sensor applications

then interfaced with a CMOS circuit implementing a sub-Nyquist ADC and other processing. The multiplier is implemented with a cross-coupled TFT based NMOS transistors and is supplied with a ± 1 random bit at 500 Hz from the CMOS circuit. Owing to the computational intensiveness of the l_1 minimization for an embedded CMOS IC, the spectral energy distribution of the EEG is extracted directly from the compressed signals using **Gabor** basis and l_2-norm minimization [19].

An example for an all-digital CS based implementation of a sensor node for a wireless bio-sensor application is shown in Fig. 7.13 [20]. Unlike the analog CS approach, the ADC (an 8-bit SAR ADC) now operates at the Nyquist rate and digitizes the incoming bio-potential signal. The ADC output drives M parallel digital paths comprising of an all-digital multiplier, an accumulator, and a sub-Nyquist sampler. The multiplication is simply a 1-bit random (± 1) signal with the 8-bit digitized signal. This can be easily implemented with an XOR gate and carry input of the accumulator (for signed addition). The accumulator is clocked into a register after every N cycles to realize the sub-Nyquist sampling and subsequently reset. At the receiver, the signal is reconstructed based on l_1-norm minimization [20].

7.8   Application to Line Spectral Estimation

Section 6.2 briefly discussed the problem of line spectral estimation (LSE), i.e., solving the parameters of the SoE model (Eq. 6.1) with $\alpha_m = 0$. The problem of line spectral estimation deals with the extraction of frequencies of the signal $\mathbf{x}$ from a small number of noisy samples of $\mathbf{y}$, where $\mathbf{y} = \mathbf{x} + \mathbf{w}$ is the measurement signal, which is the superposition of complex exponentials with additive noise $\mathbf{w}$. The signal $\mathbf{y}$ is hence sparse in the Fourier domain with finite non-zero Fourier coefficients. The vector signal $\mathbf{x}$ can be modelled as a sum of elementary waveforms called signal *atoms*. If the signal $\mathbf{x}$ is represented with T atoms, then $\mathbf{x} = \Phi\boldsymbol{\alpha} = \sum_{i=1}^{T} \alpha_i \boldsymbol{\phi}_i$, where α_i are called the decomposition coefficients of $\mathbf{x}$ in the dictionary $\Phi = [\boldsymbol{\phi}_1, \ldots, \boldsymbol{\phi}_T]$. Accurate and high resolution extraction/recovery of frequencies of the signal $\mathbf{x}$ from a finite number of quantized noisy samples of $\mathbf{y}$ is a fundamental problem in statistical signal processing and several methods with robust polynomial interpolation have been proposed. The theory of CS states that under some mild conditions on $\mathbf{A}$ (RIP), there exists a suitable convex optimization algorithm that can recover $\mathbf{x}$ using no more than $O(2k \log n)$ measurements [21, 22]. Several works have been reported in the past for applications where signal frequency lies exactly on a grid. In this discrete case, CS deals with finite dimensional and discrete frequency measurements through a discretization procedure to reduce the continuous parameter space to a finite set of grid points. However, in applications where the signal frequencies are not restricted to finite grid points, the technique does not yield high precision frequency measurements.

One way to solve the problems associated with discrete CS is to increase the number of finite grid points based on the frequency estimation accuracy required for a target application. However, finer grid points obtained by increasing the grid points are known to lead to numerical instability issues [23]. Further, in applications where the frequency does not fall onto the finite grid, the signal cannot often be sparsely represented by the discrete dictionary [23]. Additionally, increasing the number of grid points leads to the dictionary atoms being very coherent [23].

To overcome the drawbacks associated with the discrete case, analog compressed sensing based methods that exploit the general theory of atomic norms for linear inverse problems have been reported [24]. For the specific case of a sparse sum of complex sinusoids, [23, 24] provide a computationally feasible approach. These ideas are then extended in [25] for dealing with the noisy case where atomic norm soft thresholding (AST) algorithm is explored. Further, the authors in [25] also provide a reasonably fast method for solving this SDP via the alternating direction method of multipliers (ADMM). Recently, [26] proposed a low-complexity method, the Superfast Line Spectral Estimation (LSE), which draws upon ideas from Bayesian sparse estimation theory. We direct the readers to [26–28] for a detailed investigation on the superfast-LSE algorithm and as a case study present one such application for a resonant micro-electromechanical system (MEMS) sensor based particulate monitoring system (PMS), used in environmental nanoparticle (diameter < 100 nm) mass measurement for pollution monitoring [27, 29].

7.8.1 CS Based High Resolution Systems for Gravimetric MEMS Applications

Resonance based MEMS devices can be configured as gravimetric sensors wherein the mass of surface-bound species is transduced as a shift in the resonant frequency. Applications involving these bulk acoustic resonant devices rely highly on very precise measurements of the resonant frequency shift corresponding to mass change. This necessitates the need for a highly precise frequency measurement system for interfacing the bulk acoustic resonators. In environmental particle sensing applications (sub 100 nm particle mass), the bulk acoustic resonant devices are part of a sensor node. Each sensor node provides highly precise resonant frequency shifts and must cater to a range of resonant devices. For a gravimetric MEMS based particulate monitoring system (PMS), the resolution of frequency shifts ranges in a few mHz over a range of 1–10 MHz. This translates to a large resolution requirement of >20 bits with a tracking frequency range of more than 100 kHz while exposed to nanoparticles.

We now discuss the need for CS in the implementation of such a high resolution frequency measurement system. While systems based on frequency-to-digital converters (FDC) and time-to-digital converters (TDC) have been reported with a wide variety of architectures, they rely on high clock frequencies to achieve precise frequency/time measurements. For the gravimetric MEMS application, the required frequency precision leads to unrealistic high clock frequencies within a dynamic range of 100 kHz. Further, high clock frequencies are associated with high jitter in these clocks, leading to lower-resolution frequency measurements. Further, a change in the resonant mode anywhere between 1 MHz to 10 MHz implies the FDC/TDC must be re-designed either by changing the clock frequency or hardware (usually coarse counters in counter based implementations). Thus, traditional methods suffer from complex circuitry, high clock frequency requirement, jitter related resolution degradation, and non-reconfigurability.

CS has been used in the past to demonstrate that data acquisition and compression can often be combined, reducing the time and space needed to acquire signals [21, 22] by exploiting the inherent sparsity in the signal. Compressed sensing algorithms utilized at the receiver can then estimate the frequency to a high precision with a minimal data. This ensures that the ADC at the transmitter can operate at

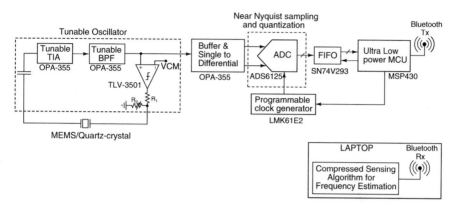

Fig. 7.14 System level description of the CS based Frequency measurement system

a minimal sampling frequency (near Nyquist over the frequency band) and with lower quantization bits while still achieving very high resolution frequency shift measurements. We discuss one such high resolution frequency measurement system in the subsequent section.

The frequency measurement system is built around the resonant MEMS based gravimetric sensor. The resonant device is incorporated in an oscillator loop whose voltage output is sampled by a moderate resolution ADC at near Nyquist sampling frequency. The complete frequency measurement system is shown in Fig. 7.14. The system has a tunable oscillator to support a range of resonant devices from 1 to 10 MHz. The tunable oscillator comprises of a trans-impedance amplifier (TIA) with tunable gain, a multi-feedback tunable BPF, and a comparator to drive the resonant MEMS device. The TIA and the BPF gain provides control on the SNR of the system. The resistive divider at the output of comparator provides control over the resonant device drive voltage level. The output of the oscillator is digitized by a 12 bit ADC ADS6125. The sampling clock for the ADC is generated by a programmable clock generator LMK61E2. A 12 bit wide and 64 kB depth FIFO is used to hold the data from the ADC. An ultra-low-power microcontroller MSP430 is used to program the clock during start-up and for data transmission through Bluetooth. Finally, a laptop is used as a receiver and the superfast-LSE CS algorithm is executed locally on the laptop for high resolution frequency estimation.

The oscillator output voltage is sampled by the ADC at a sampling rate of f_s Hz. We then use N samples to estimate one frequency value. If f_i corresponds to ith estimated frequency, then the relative error in frequency is estimated as

$$\frac{\Delta f_i}{f}(ppm) = \frac{f_{mean} - f_i}{f_{mean}} \cdot 10^6 \qquad (7.10)$$

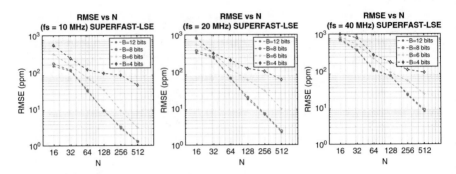

Fig. 7.15 Experimental results of RMSE vs. N plot for Superfast-LSE for three different f_s for a 4 MHz quartz crystal

The root mean square error (RMSE) is defined as standard deviation of $\frac{\Delta f_i}{f}$. Further, the SNR is statistically obtained as follows:

$$SNR = 10 \, log \frac{\displaystyle\sum_{i=1}^{i=N}(f_i)^2}{\displaystyle\sum_{i=1}^{i=N}(f_i - f_{mean})^2} \approx 10 \, log \left(\frac{10^{12}}{(RMSE_{\frac{\Delta f}{f} ppm})^2} \right) \quad (7.11)$$

With a 4 MHz quartz crystal as the resonator, Fig. 7.15 shows the RMSE of estimated frequencies for different sample lengths at three different sampling frequencies of 10 MHz, 20 MHz, and 40 MHz (for four different ADC quantization levels). The following observations can be made from the plot: (a) near Nyquist sampling provides efficient frequency estimation, which is inline with the theory of CS based LSE, (b) for target RMSE of < 100 ppm, an ADC with just 4-bit resolution suffices, thereby reducing the data size and hence the transmitter power, and (c) higher sample size yields better RMSE performance which is also inline with the theory of LSE. Based on these observations, the system sampling frequency is set to 10 MHz and for an 8-bit resolution ADC.

Figure 7.16 shows the measured frequency from the system for a MEMS sensor with a resonant frequency of 3.13 MHz. The frequency is measured for 3 h and the RMSE and the achievable SNR are evaluated. Similar measurements are performed for two different quartz crystals with resonant modes at 4 MHz and 8 MHz and another MEMS with a resonant mode at 3.02 MHz.

Table 7.1 tabulates the measured RMSE and the achievable SNR with two quartz crystal based and two MEMS based resonators, for two different sample lengths of 1536 and 8192 with an 8-bit ADC operating at 10 MHz sampling frequency. It must be noted that for the quartz crystals, which are relatively independent of temperature and humidity effects compared to a MEMS resonator, RMSE as low as 0.1 ppm can be achieved with CS based LSE. While traditional FDC circuits

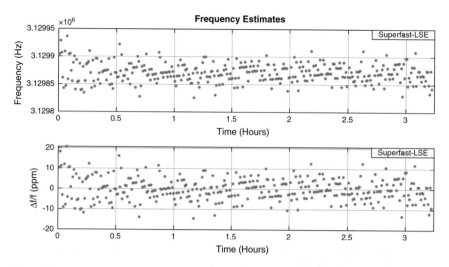

Fig. 7.16 Measured frequency from Superfast LSE for the MEMS 3.13 MHz mode with fs = 10 MHz, N = 8192, and 8 bit resolution ADC (f_{mean} = 3129871.409181 Hz, RMSE = 5.57323 ppm, SNR = 105.1 dB)

Table 7.1 Measured SNR and RMSE for two different Quartz crystals and two different MEMS modes with sample size of 1536 and 8192 samples

	$N = 1536$		$N = 8192$	
	RMSE (ppm)	SNR (dB)	RMSE (ppm)	SNR (dB)
Quartz crystal—4 MHz	1.0007	120.01	0.1078	139.35
Quartz crystal—8 MHz	0.995	120.4	0.1103	139.15
MEMS—3.02 MHz	–	–	5.7413	104.82
MEMS—3.13 MHz	–	–	5.5732	105.1

have not been reported to achieve such low RMSE or SNR (as high as 139 dB) for such applications even with complex circuit architectures and high sampling frequencies, CS based approach achieves this with just near-Nyquist sampling and a very moderate resolution ADC.

7.9 Latest Trends and Looking Ahead

Motivated by the success of deep learning, a new viewpoint has emerged for compressive sensing *model* in terms of imposing a more generic sparsity or signal information complexity on the unknown signal [30, 31]. Further, for many of the iterative methods, *algorithm unrolling* is seen as a deep neural network mapping the observed data to the signal [32–34]. Refraining ourselves from being completely exhaustive in this context, we only point out some more recent set of papers, viz.,

[35–38]. It will be an interesting research direction to investigate whether these developments can offer a *complete end-2-end* hardware implementation of sensing, as covered here, integrated with hardware execution of deep networks, thereby achieving the best overall efficiency for such systems.

References

1. S. Foucart, H. Rauhut, A mathematical introduction to compressive sensing. Bull. Am. Math **54**(2017), 151–165 (2017)
2. E.J. Candes, T. Tao, Decoding by linear programming. IEEE Trans. Inf. Theory **51**(12), 4203–4215 (2005)
3. M. Stojnic, W. Xu, B. Hassibi, Compressed sensing – probabilistic analysis of a null-space characterization, in *2008 IEEE International Conference on Acoustics, Speech and Signal Processing* (2008), pp. 3377–3380
4. J.A. Tropp, Just relax: convex programming methods for identifying sparse signals in noise. IEEE Trans. Inf. Theory **52**(3), 1030–1051 (2006)
5. E.J. Candes, Y. Plan, A probabilistic and RIPless theory of compressed sensing. IEEE Trans. Inf. Theory **57**(11), 7235–7254 (2011)
6. R. Vershynin, Estimation in high dimensions: a geometric perspective. arXiv e-prints, arXiv:1405.5103 (2014)
7. R. Vershynin, High-dimensional probability (2019). https://www.math.uci.edu/~rvershyn/papers/HDP-book/HDP-book.pdf
8. V. Chandrasekaran, B. Recht, P.A. Parrilo, A.S. Willsky, The convex geometry of linear inverse problems. Found. Comput. Math. **12**(6), 805–849 (2012)
9. S. Oymak, J.A. Tropp, Universality laws for randomized dimension reduction, with applications. Inf. Infer. A J. IMA **7**(3), 337–446 (2018)
10. S. Nam, M. Davies, M. Elad, R. Gribonval, The cosparse analysis model and algorithms. Appl. Comput. Harmonic Anal. **34**(1), 30–56 (2013). https://www.sciencedirect.com/science/article/pii/S1063520312000450
11. M. Kabanava, H. Rauhut, Cosparsity in compressed sensing, in *Compressed Sensing and Its Applications* (Springer, Berlin, 2015), pp. 315–339
12. A. Beck, M. Teboulle, A fast iterative shrinkage-thresholding algorithm for linear inverse problems. SIAM J. Imag. Sci. **2**(1), 183–202 (2009)
13. T. Zhang, Sparse recovery with orthogonal matching pursuit under rip. IEEE Trans. Inf. Theory **57**(9), 6215–6221 (2011)
14. D. Needell, J.A. Tropp, CoSaMP: iterative signal recovery from incomplete and inaccurate samples. Appl. Comput. Harmonic Anal. **26**(3), 301–321 (2009)
15. J.A. Tropp, J.N. Laska, M.F. Duarte, J.K. Romberg, R.G. Baraniuk, Beyond nyquist: efficient sampling of sparse bandlimited signals. IEEE Trans. Inf. Theory **56**(1), 520–544 (2010)
16. M. Mishali, Y.C. Eldar, From theory to practice: sub-Nyquist sampling of sparse wideband analog signals. IEEE J. Sel. Top. Sign. Proces. **4**(2), 375–391 (2010)
17. D. Gangopadhyay, E.G. Allstot, A.M.R. Dixon, K. Natarajan, S. Gupta, D.J. Allstot, Compressed sensing analog front-end for bio-sensor applications. IEEE J. Solid State Circuits **49**(2), 426–438 (2014)
18. L. Kuo, C. Hou, M. Wu, Y. Shu, A 1v 9pa analog front end with compressed sensing for electrocardiogram monitoring, in *2015 IEEE Asian Solid-State Circuits Conference (A-SSCC)* (2015), pp. 1–4
19. T. Moy, L. Huang, W. Rieutort-Louis, C. Wu, P. Cuff, S. Wagner, J.C. Sturm, N. Verma, An EEG acquisition and biomarker-extraction system using low-noise-amplifier and compressive-sensing circuits based on flexible, thin-film electronics. IEEE J. Solid State Circuits **52**(1),

309–321 (2017)

20. F. Chen, A.P. Chandrakasan, V.M. Stojanovic, Design and analysis of a hardware-efficient compressed sensing architecture for data compression in wireless sensors. IEEE J. Solid State Circuits **47**(3), 744–756 (2012)

21. E.J. Candes, J. Romberg, T. Tao, Robust uncertainty principles: exact signal reconstruction from highly incomplete frequency information. IEEE Trans. Inf. Theory **52**(2), 489–509 (2006)

22. E.J. Candes, M.B. Wakin, An introduction to compressive sampling. IEEE Signal Process. Mag. **25**(2), 21–30 (2008)

23. G. Tang, B.N. Bhaskar, P. Shah, B. Recht, Compressed sensing off the grid. IEEE Trans. Inf. Theory **59**(11), 7465–7490 (2013)

24. V. Chandrasekaran, B. Recht, P. Parrilo, A. Willsky, The convex geometry of linear inverse problems. Found. Comput. Math. **12**, 805–849 (2012). https://doi.org/10.1007/s10208-012-9135-7

25. B.N. Bhaskar, G. Tang, B. Recht, Atomic norm denoising with applications to line spectral estimation. IEEE Trans. Signal Process. **61**(23), 5987–5999 (2013)

26. T.L. Hansen, B.H. Fleury, B.D. Rao, Superfast line spectral estimation. IEEE Trans. Signal Process. **66**(10), 2511–2526 (2018)

27. L. Somappa, S. Aeron, A.G. Menon, S. Sonkusale, A.A. Seshia, M.S. Baghini, On quantized analog compressive sensing methods for efficient resonator frequency estimation. IEEE Trans. Circuits Syst. I Regl. Pap. **67**(12), 4556–4565 (2020)

28. L. Somappa, S. Malik, S. Aeron, S. Sonkusale, M.S. Baghini, High resolution frequency measurement techniques for relaxation oscillator based capacitive sensors. IEEE Sensors J. 1–1 (2021)

29. M. Chellasivalingam, L. Somappa, A.M. Boies, M.S. Baghini, A.A. Seshia, MEMS based gravimetric sensor for the detection of ultra-fine aerosol particles, in *2020 IEEE SENSORS* (2020), pp. 1–4

30. A. Bora, A. Jalal, E. Price, A.G. Dimakis, Compressed sensing using generative models, in *International Conference on Machine Learning* (PMLR, Cambridge, 2017), pp. 537–546

31. V. Lempitsky, A. Vedaldi, D. Ulyanov, Deep image prior, in *2018 IEEE/CVF Conference on Computer Vision and Pattern Recognition* (IEEE, Piscataway, 2018), pp. 9446–9454

32. D. Ito, S. Takabe, T. Wadayama, Trainable ISTA for sparse signal recovery. IEEE Trans. Signal Process. **67**(12), 3113–3125 (2019)

33. U.S. Kamilov, H. Mansour, Learning optimal nonlinearities for iterative thresholding algorithms. IEEE Signal Process. Lett. **23**(5), 747–751 (2016)

34. M. Mardani, Q. Sun, S. Vasawanala, V. Papyan, H. Monajemi, J. Pauly, D. Donoho, Neural proximal gradient descent for compressive imaging (2018). arXiv. Preprint. arXiv:1806.03963

35. A. Mousavi, R.G. Baraniuk, Learning to invert: Signal recovery via deep convolutional networks, in *2017 IEEE International Conference on Acoustics, Speech and Signal Processing (ICASSP)* (IEEE, Piscataway, 2017), pp. 2272–2276

36. Y. Wu, M. Rosca, T. Lillicrap, Deep compressed sensing, in *International Conference on Machine Learning* (PMLR, Cambridge, 2019), pp. 6850–6860

37. S. Khobahi, M. Soltanalian, Model-aware deep architectures for one-bit compressive variational autoencoding. arXiv e-prints, pp. arXiv–1911 (2019)

38. Z. Zhang, Y. Liu, J. Liu, F. Wen, C. Zhu, Amp-net: denoising-based deep unfolding for compressive image sensing. IEEE Trans. Image Process. **30**, 1487–1500 (2021)

Index

© The Author(s), under exclusive license to Springer Nature Switzerland AG 2022
S. Sonkusale et al., *Flexible Bioelectronics with Power Autonomous Sensing
and Data Analytics*, https://doi.org/10.1007/978-3-030-98538-7

Printed in the United States
by Baker & Taylor Publisher Services